Everyone's Guide to Planet Jupiter

Compiled by

Roselle Garland

Scribbles

Year of Publication 2018

ISBN : 9789352979530

Book Published by

Scribbles

(An Imprint of Alpha Editions)

email - alphaedis@gmail.com

Produced by: PediaPress GmbH
Limburg an der Lahn
Germany
http://pediapress.com/

Contents

Introduction

Jupiter

<indicator name="pp-default"> 🔒 </indicator> <indicator name="featured-star"> ⭐ </indicator>

Jupiter

Full-disc view of Jupiter in natural color in April 2014[1]

Designations	
Pronunciation	/ˈdʒuːpɪtər/ (🔊 listen)
Adjectives	Jovian
Orbital characteristics	
Epoch J2000	
Aphelion	816.62 million km (5.4588 AU)
Perihelion	740.52 million km (4.9501 AU)
Semi-major axis	778.57 million km (5.2044 AU)
Eccentricity	0.0489

Orbital period	• 11.862 yr • 4,332.59 d • 10,475.8 Jovian solar days
Synodic period	398.88 d
Average orbital speed	13.07 km/s (8.12 mi/s)
Mean anomaly	20.020°
Inclination	• 1.303° to ecliptic • 6.09° to Sun's equator • 0.32° to invariable plane
Longitude of ascending node	100.464°
Argument of perihelion	273.867°
Known satellites	79 (as of 2018[2])
Physical characteristics	
Mean radius	69,911 km (43,441 mi)
Equatorial radius	• 71,492 km (44,423 mi)[3] • 11.209 Earths
Polar radius	• 66,854 km (41,541 mi) • 10.517 Earths
Flattening	0.06487
Surface area	• 6.1419×10^{10} km^2 (2.3714×10^{10} sq mi) • 121.9 Earths
Volume	• 1.4313×10^{15} km^3 (3.434×10^{14} cu mi) • 1,321 Earths
Mass	• 1.8982×10^{27} kg (4.1848×10^{27} lb) • 317.8 Earths • 1/1047 Sun
Mean density	1,326 kg/m^3 (2,235 lb/cu yd)[4]
Surface gravity	24.79 m/s^2 (81.3 ft/s^2) 2.528 g
Moment of inertia factor	0.254 I/MR^2 (estimate)
Escape velocity	59.5 km/s (37.0 mi/s)
Sidereal rotation period	9.925 hours (9 h 55 m 30 s)
Equatorial rotation velocity	12.6 km/s (7.8 mi/s; 45,000 km/h)
Axial tilt	3.13° (to orbit)
North pole right ascension	268.057°; 17^h 52^m 14^s
North pole declination	64.495°

Albedo	0.343 (Bond) 0.538 (geometric)			

	Surface temp.	min	mean	max
	1 bar level		165 K (−108 °C)	
	0.1 bar		112 K (−161 °C)	

Apparent magnitude	−1.6 to −2.94
Angular diameter	29.8" to 50.1"

Atmosphere	
Surface pressure	20–200 kPa; 70 kPa
Scale height	27 km (17 mi)
Composition by volume	by volume:

	89%±2.0%	hydrogen (H_2)
	10%±2.0%	helium (He)
	0.3%±0.1%	methane (CH_4)
	0.026%±0.004%	ammonia (NH_3)
	0.0028%±0.001%	hydrogen deuteride (HD)
	0.0006%±0.0002%	ethane (C_2H_6)
	0.0004%±0.0004%	water (H_2O)

Ices:
- ammonia (NH_3)
- water (H_2O)
- ammonium hydrosulfide (NH_4SH)

Jupiter is the fifth planet from the Sun and the largest in the Solar System. It is a giant planet with a mass one-thousandth that of the Sun, but two-and-a-half times that of all the other planets in the Solar System combined. Jupiter and Saturn are gas giants; the other two giant planets, Uranus and Neptune, are ice giants. Jupiter has been known to astronomers since antiquity. The Romans named it after their god Jupiter. When viewed from Earth, Jupiter can reach an apparent magnitude of −2.94, bright enough for its reflected light to cast shadows, and making it on average the third-brightest natural object in the night sky after the Moon and Venus.

Jupiter is primarily composed of hydrogen with a quarter of its mass being helium, though helium comprises only about a tenth of the number of molecules.

It may also have a rocky core of heavier elements, but like the other giant planets, Jupiter lacks a well-defined solid surface. Because of its rapid rotation, the planet's shape is that of an oblate spheroid (it has a slight but noticeable bulge around the equator). The outer atmosphere is visibly segregated into several bands at different latitudes, resulting in turbulence and storms along their interacting boundaries. A prominent result is the Great Red Spot, a giant storm that is known to have existed since at least the 17th century when it was first seen by telescope. Surrounding Jupiter is a faint planetary ring system and a powerful magnetosphere. Jupiter has at least 79 moons, including the four large Galilean moons discovered by Galileo Galilei in 1610. Ganymede, the largest of these, has a diameter greater than that of the planet Mercury.

Jupiter has been explored on several occasions by robotic spacecraft, most notably during the early *Pioneer* and *Voyager* flyby missions and later by the *Galileo* orbiter. In late February 2007, Jupiter was visited by the *New Horizons* probe, which used Jupiter's gravity to increase its speed and bend its trajectory en route to Pluto. The latest probe to visit the planet is *Juno*, which entered into orbit around Jupiter on July 4, 2016. Future targets for exploration in the Jupiter system include the probable ice-covered liquid ocean of its moon Europa.

Formation and migration

Earth and its neighbor planets may have formed from fragments of planets after collisions with Jupiter destroyed those super-Earths near the Sun. As Jupiter came toward the inner Solar System, in what theorists call the grand tack hypothesis, gravitational tugs and pulls occurred causing a series of collisions between the super-Earths as their orbits began to overlap.

Astronomers have discovered nearly 500 planetary systems with multiple planets. Regularly these systems include a few planets with masses several times greater than Earth's (super-Earths), orbiting closer to their star than Mercury is to the Sun, and sometimes also Jupiter-mass gas giants close to their star.

Jupiter moving out of the inner Solar System would have allowed the formation of inner planets, including Earth.

Physical characteristics

Jupiter is composed primarily of gaseous and liquid matter. It is the largest of the four giant planets in the Solar System and hence its largest planet. It has a diameter of 142,984 km (88,846 mi) at its equator. The average density of Jupiter, 1.326 g/cm^3, is the second highest of the giant planets, but lower than those of the four terrestrial planets.

Composition

Jupiter's upper atmosphere is about 88–92% hydrogen and 8–12% helium by percent volume of gas molecules. A helium atom has about four times as much mass as a hydrogen atom, so the composition changes when described as the proportion of mass contributed by different atoms. Thus, Jupiter's atmosphere is approximately 75% hydrogen and 24% helium by mass, with the remaining one percent of the mass consisting of other elements. The atmosphere contains trace amounts of methane, water vapor, ammonia, and silicon-based compounds. There are also traces of carbon, ethane, hydrogen sulfide, neon, oxygen, phosphine, and sulfur. The outermost layer of the atmosphere contains crystals of frozen ammonia. The interior contains denser materials - by mass it is roughly 71% hydrogen, 24% helium, and 5% other elements. Through infrared and ultraviolet measurements, trace amounts of benzene and other hydrocarbons have also been found.

The atmospheric proportions of hydrogen and helium are close to the theoretical composition of the primordial solar nebula. Neon in the upper atmosphere only consists of 20 parts per million by mass, which is about a tenth as abundant as in the Sun. Helium is also depleted to about 80% of the Sun's helium composition. This depletion is a result of precipitation of these elements into the interior of the planet.

Based on spectroscopy, Saturn is thought to be similar in composition to Jupiter, but the other giant planets Uranus and Neptune have relatively less hydrogen and helium and relatively more ices and are thus now termed ice giants.

Mass and size

Jupiter's mass is 2.5 times that of all the other planets in the Solar System combined—this is so massive that its barycenter with the Sun lies above the Sun's surface at 1.068 solar radii from the Sun's center. Jupiter is much larger than Earth and considerably less dense: its volume is that of about 1,321 Earths, but it is only 318 times as massive. Jupiter's radius is about 1/10 the radius of the Sun, and its mass is 0.001 times the mass of the Sun, so the densities of the two bodies are similar. A "Jupiter mass" (M_J or M_{Jup}) is often used as a unit to describe masses of other objects, particularly extrasolar planets and brown dwarfs. So, for example, the extrasolar planet HD 209458 b has a mass of 0.69 M_J, while Kappa Andromedae b has a mass of 12.8 M_J.

Theoretical models indicate that if Jupiter had much more mass than it does at present, it would shrink. For small changes in mass, the radius would not change appreciably, and above about 500 $M_\oplus$ (1.6 Jupiter masses) the interior would become so much more compressed under the increased pressure that its

Figure 1: *Jupiter's diameter is one order of magnitude smaller ($\times 0.10045$)*
than that of the Sun, and one order of magnitude larger ($\times 10.9733$)
than that of Earth. The Great Red Spot is roughly the same size as Earth.

volume would *decrease* despite the increasing amount of matter. As a result, Jupiter is thought to have about as large a diameter as a planet of its composition and evolutionary history can achieve. The process of further shrinkage with increasing mass would continue until appreciable stellar ignition was achieved, as in high-mass brown dwarfs having around 50 Jupiter masses.

Although Jupiter would need to be about 75 times as massive to fuse hydrogen and become a star, the smallest red dwarf is only about 30 percent larger in radius than Jupiter. Despite this, Jupiter still radiates more heat than it receives from the Sun; the amount of heat produced inside it is similar to the total solar radiation it receives. This additional heat is generated by the Kelvin–Helmholtz mechanism through contraction. This process causes Jupiter to shrink by about 2 cm each year. When it was first formed, Jupiter was much hotter and was about twice its current diameter.

Figure 2: *Animation of four images showing Jupiter in infrared light as seen by NASA's Infrared telescope facility on May 16, 2015*

Internal structure

Jupiter is thought to consist of a dense core with a mixture of elements, a surrounding layer of liquid metallic hydrogen with some helium, and an outer layer predominantly of molecular hydrogen. Beyond this basic outline, there is still considerable uncertainty. The core is often described as rocky, but its detailed composition is unknown, as are the properties of materials at the temperatures and pressures of those depths (see below). In 1997, the existence of the core was suggested by gravitational measurements, indicating a mass of from 12 to 45 times that of Earth, or roughly 4%–14% of the total mass of Jupiter. The presence of a core during at least part of Jupiter's history is suggested by models of planetary formation that require the formation of a rocky or icy core massive enough to collect its bulk of hydrogen and helium from the protosolar nebula. Assuming it did exist, it may have shrunk as convection currents of hot liquid metallic hydrogen mixed with the molten core and carried its contents to higher levels in the planetary interior. A core may now be entirely absent, as gravitational measurements are not yet precise enough to rule that possibility out entirely.

The uncertainty of the models is tied to the error margin in hitherto measured parameters: one of the rotational coefficients (J_6) used to describe the planet's gravitational moment, Jupiter's equatorial radius, and its temperature at 1 bar pressure. The *Juno* mission, which arrived in July 2016, is expected to further constrain the values of these parameters for better models of the core.

The core region may be surrounded by dense metallic hydrogen, which extends outward to about 78% of the radius of the planet. Rain-like droplets of helium and neon precipitate downward through this layer, depleting the abundance of these elements in the upper atmosphere. Rainfalls of diamonds have been suggested to occur on Jupiter, as well as on Saturn and ice giants Uranus and Neptune.

Above the layer of metallic hydrogen lies a transparent interior atmosphere of hydrogen. At this depth, the pressure and temperature are above hydrogen's critical pressure of 1.2858 MPa and critical temperature of only 32.938 K. In this state, there are no distinct liquid and gas phases—hydrogen is said to be in a supercritical fluid state. It is convenient to treat hydrogen as gas in the upper layer extending downward from the cloud layer to a depth of about 1,000 km, and as liquid in deeper layers. Physically, there is no clear boundary—the gas smoothly becomes hotter and denser as one descends.

The temperature and pressure inside Jupiter increase steadily toward the core, due to the Kelvin–Helmholtz mechanism. At the pressure level of 10 bars (1 MPa), the temperature is around 340 K (67 °C; 152 °F). At the phase transition region where hydrogen—heated beyond its critical point—becomes metallic, it is calculated the temperature is 10,000 K (9,700 °C; 17,500 °F) and the pressure is 200 GPa. The temperature at the core boundary is estimated to be 36,000 K (35,700 °C; 64,300 °F) and the interior pressure is roughly 3,000–4,500 GPa.

Atmosphere

Jupiter has the largest planetary atmosphere in the Solar System, spanning over 5,000 km (3,000 mi) in altitude. Because Jupiter has no surface, the base of its atmosphere is usually considered to be the point at which atmospheric pressure is equal to 100 kPa (1.0 bar).

Cloud layers

<templatestyles src="Multiple_image/styles.css" />

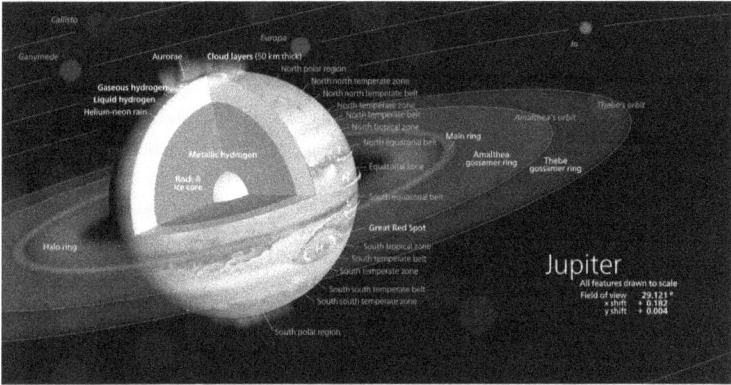

Figure 3: *This cut-away illustrates a model of the interior of Jupiter, with a rocky core overlaid by a deep layer of liquid metallic hydrogen.*

Figure 4: *The movement of Jupiter's counter-rotating cloud bands. This looping animation maps the planet's exterior onto a cylindrical projection.*

South polar view of Jupiter

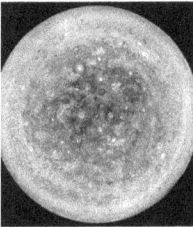

Enhanced color view of Jupiter's southern storms

Jupiter is perpetually covered with clouds composed of ammonia crystals and possibly ammonium hydrosulfide. The clouds are located in the tropopause and are arranged into bands of different latitudes, known as tropical regions.

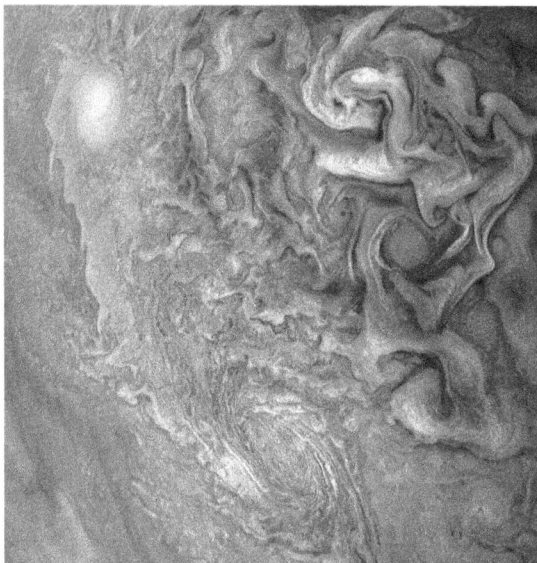

Figure 5: *Jupiter clouds*
(Juno; December 2017)

These are sub-divided into lighter-hued *zones* and darker *belts*. The inter-actions of these conflicting circulation patterns cause storms and turbulence. Wind speeds of 100 m/s (360 km/h) are common in zonal jets. The zones have been observed to vary in width, color and intensity from year to year, but they have remained sufficiently stable for scientists to give them identifying designations.

The cloud layer is only about 50 km (31 mi) deep, and consists of at least two decks of clouds: a thick lower deck and a thin clearer region. There may also be a thin layer of water clouds underlying the ammonia layer. Supporting the idea of water clouds are the flashes of lightning detected in the atmosphere of Jupiter. These electrical discharges can be up to a thousand times as powerful as lightning on Earth. The water clouds are assumed to generate thunderstorms in the same way as terrestrial thunderstorms, driven by the heat rising from the interior.

The orange and brown coloration in the clouds of Jupiter are caused by up-welling compounds that change color when they are exposed to ultraviolet light from the Sun. The exact makeup remains uncertain, but the substances are thought to be phosphorus, sulfur or possibly hydrocarbons. These color-ful compounds, known as chromophores, mix with the warmer, lower deck of

Figure 6: *Time-lapse sequence from the approach of Voyager 1, showing the motion of atmospheric bands and circulation of the Great Red Spot. Recorded over 32 days with one photograph taken every 10 hours (once per Jovian day). See full size video.*

clouds. The zones are formed when rising convection cells form crystallizing ammonia that masks out these lower clouds from view.

Jupiter's low axial tilt means that the poles constantly receive less solar radiation than at the planet's equatorial region. Convection within the interior of the planet transports more energy to the poles, balancing out the temperatures at the cloud layer.

Great Red Spot and other vortices

The best known feature of Jupiter is the Great Red Spot, a persistent anticyclonic storm that is larger than Earth, located 22° south of the equator. It is known to have been in existence since at least 1831, and possibly since 1665.[5] Images by the Hubble Space Telescope have shown as many as two "red spots" adjacent to the Great Red Spot. The storm is large enough to be visible through Earth-based telescopes with an aperture of 12 cm or larger. The oval object rotates counterclockwise, with a period of about six days. The maximum altitude of this storm is about 8 km (5 mi) above the surrounding cloudtops.

The Great Red Spot is large enough to accommodate Earth within its boundaries. Mathematical models suggest that the storm is stable and may be a

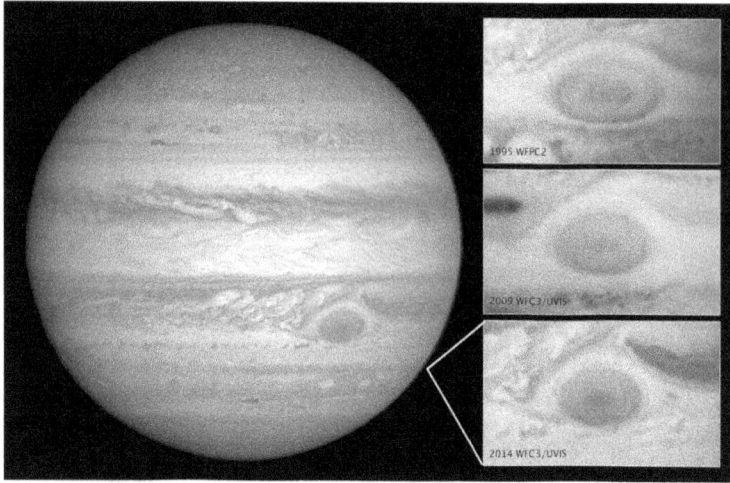

Figure 7: *Great Red Spot is decreasing in size (May 15, 2014).*

permanent feature of the planet. However, it has significantly decreased in size since its discovery. Initial observations in the late 1800s showed it to be approximately 41,000 km (25,500 mi) across. By the time of the *Voyager* fly-bys in 1979, the storm had a length of 23,300 km (14,500 mi) and a width of approximately 13,000 km (8,000 mi). *Hubble* observations in 1995 showed it had decreased in size again to 20,950 km (13,020 mi), and observations in 2009 showed the size to be 17,910 km (11,130 mi). As of 2015[2], the storm was measured at approximately 16,500 by 10,940 km (10,250 by 6,800 mi), and is decreasing in length by about 930 km (580 mi) per year.

Storms such as this are common within the turbulent atmospheres of giant planets. Jupiter also has white ovals and brown ovals, which are lesser un-named storms. White ovals tend to consist of relatively cool clouds within the upper atmosphere. Brown ovals are warmer and located within the "normal cloud layer". Such storms can last as little as a few hours or stretch on for centuries.

Even before Voyager proved that the feature was a storm, there was strong evidence that the spot could not be associated with any deeper feature on the planet's surface, as the Spot rotates differentially with respect to the rest of the atmosphere, sometimes faster and sometimes more slowly.

In 2000, an atmospheric feature formed in the southern hemisphere that is similar in appearance to the Great Red Spot, but smaller. This was created when several smaller, white oval-shaped storms merged to form a single fea-ture—these three smaller white ovals were first observed in 1938. The merged feature was named Oval BA, and has been nicknamed Red Spot Junior. It has since increased in intensity and changed color from white to red.

In April 2017, scientists reported the discovery of a "Great Cold Spot" in Jupiter's thermosphere at its north pole that is 24,000 km (15,000 mi) across, 12,000 km (7,500 mi) wide, and 200 °C (360 °F) cooler than surrounding material. The feature was discovered by researchers at the Very Large Telescope in Chile, who then searched archived data from the NASA Infrared Telescope Facility between 1995 and 2000. They found that, while the Spot changes size, shape and intensity over the short term, it has maintained its general position in the atmosphere across more than 15 years of available data. Scientists believe the Spot is a giant vortex similar to the Great Red Spot and also appears to be quasi-stable like the vortices in Earth's thermosphere. Interactions between charged particles generated from Io and the planet's strong magnetic field likely resulted in redistribution of heat flow, forming the Spot.

Magnetosphere

<templatestyles src="Multiple_image/styles.css" />

Aurorae on the north pole of Jupiter as viewed by Hubble

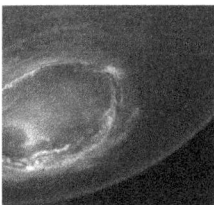

Infrared view of Jupiter's southern lights, taken by the Jovian Infrared Auroral Mapper

Jupiter's magnetic field is fourteen times as strong as that of Earth, ranging from 4.2 gauss (0.42 mT) at the equator to 10–14 gauss (1.0–1.4 mT) at the poles, making it the strongest in the Solar System (except for sunspots). This field is thought to be generated by eddy currents—swirling movements of conducting materials—within the liquid metallic hydrogen core. The volcanoes on the moon Io emit large amounts of sulfur dioxide forming a gas torus along the moon's orbit. The gas is ionized in the magnetosphere producing sulfur and oxygen ions. They, together with hydrogen ions originating from the atmosphere of Jupiter, form a plasma sheet in Jupiter's equatorial plane. The plasma in the sheet co-rotates with the planet causing deformation of the dipole magnetic field into that of magnetodisk. Electrons within the plasma sheet generate a strong radio signature that produces bursts in the range of 0.6–30 MHz.

At about 75 Jupiter radii from the planet, the interaction of the magnetosphere with the solar wind generates a bow shock. Surrounding Jupiter's magnetosphere is a magnetopause, located at the inner edge of a magnetosheath—a region between it and the bow shock. The solar wind interacts with these regions, elongating the magnetosphere on Jupiter's lee side and extending it outward until it nearly reaches the orbit of Saturn. The four largest moons of Jupiter all orbit within the magnetosphere, which protects them from the solar wind.

The magnetosphere of Jupiter is responsible for intense episodes of radio emission from the planet's polar regions. Volcanic activity on Jupiter's moon Io (see below) injects gas into Jupiter's magnetosphere, producing a torus of particles about the planet. As Io moves through this torus, the interaction generates Alfvén waves that carry ionized matter into the polar regions of Jupiter. As a result, radio waves are generated through a cyclotron maser mechanism, and the energy is transmitted out along a cone-shaped surface. When Earth intersects this cone, the radio emissions from Jupiter can exceed the solar radio output.

Orbit and rotation

Jupiter is the only planet whose barycenter with the Sun lies outside the volume of the Sun, though by only 7% of the Sun's radius.[6] The average distance between Jupiter and the Sun is 778 million km (about 5.2 times the average distance between Earth and the Sun, or 5.2 AU) and it completes an orbit every 11.86 years. This is approximately two-fifths the orbital period of Saturn, forming a near orbital resonance between the two largest planets in the Solar System. The elliptical orbit of Jupiter is inclined 1.31° compared to Earth. Because the eccentricity of its orbit is 0.048, Jupiter's distance from the Sun varies by 75 million km between its nearest approach (perihelion) and furthest distance (aphelion).

The axial tilt of Jupiter is relatively small: only 3.13°. As a result, it does not experience significant seasonal changes, in contrast to, for example, Earth and Mars.

Jupiter's rotation is the fastest of all the Solar System's planets, completing a rotation on its axis in slightly less than ten hours; this creates an equatorial bulge

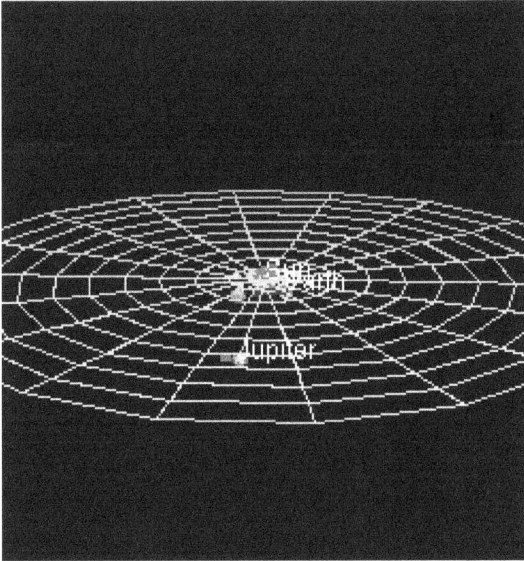

Figure 8: *Jupiter (red) completes one orbit of the Sun (center) for every 11.86 orbits of Earth (blue)*

easily seen through an Earth-based amateur telescope. The planet is shaped as an oblate spheroid, meaning that the diameter across its equator is longer than the diameter measured between its poles. On Jupiter, the equatorial diameter is 9,275 km (5,763 mi) longer than the diameter measured through the poles.

Because Jupiter is not a solid body, its upper atmosphere undergoes differential rotation. The rotation of Jupiter's polar atmosphere is about 5 minutes longer than that of the equatorial atmosphere; three systems are used as frames of reference, particularly when graphing the motion of atmospheric features. System I applies from the latitudes 10° N to 10° S; its period is the planet's shortest, at 9h 50m 30.0s. System II applies at all latitudes north and south of these; its period is 9h 55m 40.6s. System III was first defined by radio astronomers, and corresponds to the rotation of the planet's magnetosphere; its period is Jupiter's official rotation.

Observation

<templatestyles src="Multiple_image/styles.css" />

Conjunction of Jupiter and the Moon

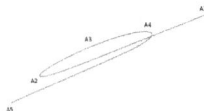

The retrograde motion of an outer planet is caused by its relative location with respect to Earth

Jupiter is usually the fourth brightest object in the sky (after the Sun, the Moon and Venus); at times Mars appears brighter than Jupiter. Depending on Jupiter's position with respect to the Earth, it can vary in visual magnitude from as bright as –2.9 at opposition down to –1.6 during conjunction with the Sun. The angular diameter of Jupiter likewise varies from 50.1 to 29.8 arc seconds. Favorable oppositions occur when Jupiter is passing through perihelion, an event that occurs once per orbit.

Earth overtakes Jupiter every 398.9 days as it orbits the Sun, a duration called the synodic period. As it does so, Jupiter appears to undergo retrograde motion with respect to the background stars. That is, for a period Jupiter seems to move backward in the night sky, performing a looping motion.

Because the orbit of Jupiter is outside that of Earth, the phase angle of Jupiter as viewed from Earth never exceeds 11.5°. That is, the planet always appears nearly fully illuminated when viewed through Earth-based telescopes. It was only during spacecraft missions to Jupiter that crescent views of the planet were obtained. A small telescope will usually show Jupiter's four Galilean moons and the prominent cloud belts across Jupiter's atmosphere. A large telescope will show Jupiter's Great Red Spot when it faces Earth.

Mythology

The planet Jupiter has been known since ancient times. It is visible to the naked eye in the night sky and can occasionally be seen in the daytime when the Sun is low. To the Babylonians, this object represented their god Marduk. They used Jupiter's roughly 12-year orbit along the ecliptic to define the constellations of their zodiac.

IVPITEI

Figure 9: *Jupiter, woodcut from a 1550 edition of Guido Bonatti's Liber Astronomiae*

The Romans named it after Jupiter (Latin: *Iuppiter, Iūpiter*) (also called Jove), the principal god of Roman mythology, whose name comes from the Proto-Indo-European vocative compound **Dyēu-pəter* (nominative: **Dyēus-pətēr*, meaning "Father Sky-God", or "Father Day-God"). In turn, Jupiter was the counterpart to the mythical Greek *Zeus* (Ζεύς), also referred to as *Dias* (Δίας), the planetary name of which is retained in modern Greek.[7]

The astronomical symbol for the planet, ♃ , is a stylized representation of the god's lightning bolt. The original Greek deity *Zeus* supplies the root *zeno-*, used to form some Jupiter-related words, such as *zenographic*.[8]

Jovian is the adjectival form of Jupiter. The older adjectival form *jovial*, employed by astrologers in the Middle Ages, has come to mean "happy" or "merry", moods ascribed to Jupiter's astrological influence.

The Chinese, Vietnamese, Koreans and Japanese called it the "wood star" (Chinese: 木星 ; pinyin: *mùxīng*), based on the Chinese Five Elements. Chinese Taoism personified it as the Fu star. The Greeks called it *Φαέθων* (*Phaethon*, meaning "blazing").

In Vedic astrology, Hindu astrologers named the planet after Brihaspati, the religious teacher of the gods, and often called it "Guru", which literally means the "Heavy One".

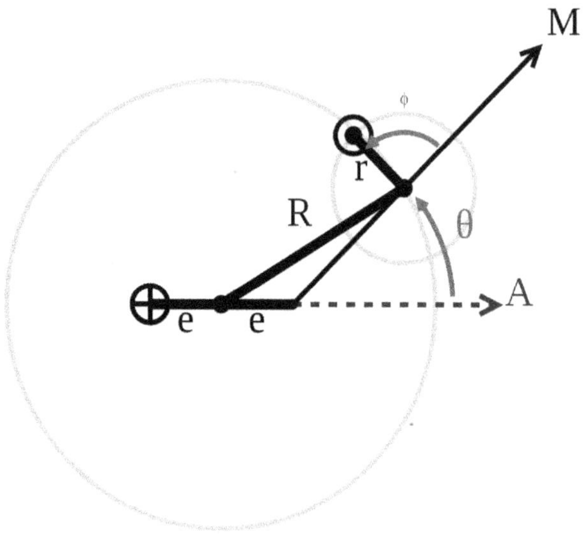

Figure 10: *Model in the Almagest of the longitu-dinal motion of Jupiter (♃) relative to Earth (⊕)*

In Germanic mythology, Jupiter is equated to Thor, whence the English name *Thursday* for the Roman *dies Jovis*.

In the Central Asian-Turkic myths, Jupiter is called *Erendiz* or *Erentüz*, from *eren* (of uncertain meaning) and *yultuz* ("star"). There are many theories about the meaning of *eren*. These peoples calculated the period of the orbit of Jupiter as 11 years and 300 days. They believed that some social and natural events connected to Erentüz's movements on the sky.

History of research and exploration

Pre-telescopic research

The observation of Jupiter dates back to at least the Babylonian astronomers of the 7th or 8th century BC. The ancient Chinese also observed the orbit of *Suìxīng* (歲星) and established their cycle of 12 earthly branches based on its approximate number of years; the Chinese language still uses its name (sim-plified as 岁) when referring to years of age. By the 4th century BC, these observations had developed into the Chinese zodiac, with each year associ-ated with a Tai Sui star and god controlling the region of the heavens oppo-site Jupiter's position in the night sky; these beliefs survive in some Taoist

Figure 11: *Galileo Galilei, discoverer of the four moons of Jupiter, now known as Galilean moons*

religious practices and in the East Asian zodiac's twelve animals, now often popularly assumed to be related to the arrival of the animals before Buddha. The Chinese historian Xi Zezong has claimed that Gan De, an ancient Chinese astronomer, discovered one of Jupiter's moons in 362 BC with the unaided eye. If accurate, this would predate Galileo's discovery by nearly two millennia. In his 2nd century work the *Almagest*, the Hellenistic astronomer Claudius Ptolemaeus constructed a geocentric planetary model based on deferents and epicycles to explain Jupiter's motion relative to Earth, giving its orbital period around Earth as 4332.38 days, or 11.86 years. In 499, Aryabhata, a mathematician–astronomer from the classical age of Indian mathematics and astronomy, also used a geocentric model to estimate Jupiter's period as 4332.2722 days, or 11.86 years.[9]Wikipedia:Verifiability

Ground-based telescope research

In 1610, Italian polymath Galileo Galilei discovered the four largest moons of Jupiter (now known as the Galilean moons) using a telescope; thought to be the first telescopic observation of moons other than Earth's. One day after Galileo, Simon Marius independently discovered moons around Jupiter, though he did not publish his discovery in a book until 1614. It was Marius's names for the four major moons, however, that stuck—Io, Europa, Ganymede and Callisto.

These findings were also the first discovery of celestial motion not apparently centered on Earth. The discovery was a major point in favor of Copernicus' heliocentric theory of the motions of the planets; Galileo's outspoken support of the Copernican theory placed him under the threat of the Inquisition.

During the 1660s, Giovanni Cassini used a new telescope to discover spots and colorful bands on Jupiter and observed that the planet appeared oblate; that is, flattened at the poles. He was also able to estimate the rotation period of the planet. In 1690 Cassini noticed that the atmosphere undergoes differential rotation.

The Great Red Spot, a prominent oval-shaped feature in the southern hemisphere of Jupiter, may have been observed as early as 1664 by Robert Hooke and in 1665 by Cassini, although this is disputed. The pharmacist Heinrich Schwabe produced the earliest known drawing to show details of the Great Red Spot in 1831.

The Red Spot was reportedly lost from sight on several occasions between 1665 and 1708 before becoming quite conspicuous in 1878. It was recorded as fading again in 1883 and at the start of the 20th century.

Both Giovanni Borelli and Cassini made careful tables of the motions of Jupiter's moons, allowing predictions of the times when the moons would pass before or behind the planet. By the 1670s, it was observed that when Jupiter was on the opposite side of the Sun from Earth, these events would occur about 17 minutes later than expected. Ole Rømer deduced that light does not travel instantaneously (a conclusion that Cassini had earlier rejected), and this timing discrepancy was used to estimate the speed of light.

In 1892, E. E. Barnard observed a fifth satellite of Jupiter with the 36-inch (910 mm) refractor at Lick Observatory in California. The discovery of this relatively small object, a testament to his keen eyesight, quickly made him famous. This moon was later named Amalthea. It was the last planetary moon to be discovered directly by visual observation.

In 1932, Rupert Wildt identified absorption bands of ammonia and methane in the spectra of Jupiter.

Three long-lived anticyclonic features termed white ovals were observed in 1938. For several decades they remained as separate features in the atmosphere, sometimes approaching each other but never merging. Finally, two of the ovals merged in 1998, then absorbed the third in 2000, becoming Oval BA.

Figure 12: *Infrared image of Jupiter taken by ESO's Very Large Telescope*

Radiotelescope research

In 1955, Bernard Burke and Kenneth Franklin detected bursts of radio signals coming from Jupiter at 22.2 MHz. The period of these bursts matched the rotation of the planet, and they were also able to use this information to refine the rotation rate. Radio bursts from Jupiter were found to come in two forms: long bursts (or L-bursts) lasting up to several seconds, and short bursts (or S-bursts) that had a duration of less than a hundredth of a second.

Scientists discovered that there were three forms of radio signals transmitted from Jupiter.

- Decametric radio bursts (with a wavelength of tens of meters) vary with the rotation of Jupiter, and are influenced by interaction of Io with Jupiter's magnetic field.
- Decimetric radio emission (with wavelengths measured in centimeters) was first observed by Frank Drake and Hein Hvatum in 1959. The origin of this signal was from a torus-shaped belt around Jupiter's equator. This signal is caused by cyclotron radiation from electrons that are accelerated in Jupiter's magnetic field.
- Thermal radiation is produced by heat in the atmosphere of Jupiter.

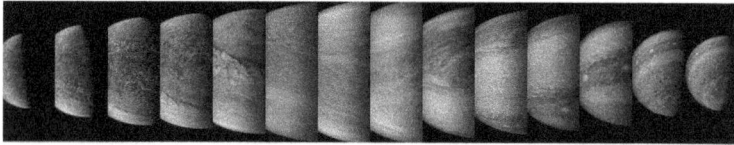

Figure 13:
Perijove 6 pass of Jupiter as viewed by JunoCam

Exploration

Since 1973 a number of automated spacecraft have visited Jupiter, most notably the *Pioneer 10* space probe, the first spacecraft to get close enough to Jupiter to send back revelations about the properties and phenomena of the Solar System's largest planet.[10,11] Flights to other planets within the Solar System are accomplished at a cost in energy, which is described by the net change in velocity of the spacecraft, or delta-v. Entering a Hohmann transfer orbit from Earth to Jupiter from low Earth orbit requires a delta-v of 6.3 km/s[12] which is comparable to the 9.7 km/s delta-v needed to reach low Earth orbit. Fortunately, gravity assists through planetary flybys can be used to reduce the energy required to reach Jupiter, albeit at the cost of a significantly longer flight duration.

Flyby missions

Flyby missions

Spacecraft	Closest approach	Distance
Pioneer 10	December 3, 1973	130,000 km
Pioneer 11	December 4, 1974	34,000 km
Voyager 1	March 5, 1979	349,000 km
Voyager 2	July 9, 1979	570,000 km
Ulysses	February 8, 1992	408,894 km
	February 4, 2004	120,000,000 km
Cassini	December 30, 2000	10,000,000 km
New Horizons	February 28, 2007	2,304,535 km

Beginning in 1973, several spacecraft have performed planetary flyby maneuvers that brought them within observation range of Jupiter. The Pioneer missions obtained the first close-up images of Jupiter's atmosphere and several of its moons. They discovered that the radiation fields near the planet were much stronger than expected, but both spacecraft managed to survive in that environment. The trajectories of these spacecraft were used to refine the mass estimates of the Jovian system. Radio occultations by the planet resulted in better measurements of Jupiter's diameter and the amount of polar flattening.

Six years later, the Voyager missions vastly improved the understanding of the Galilean moons and discovered Jupiter's rings. They also confirmed that the Great Red Spot was anticyclonic. Comparison of images showed that the Red Spot had changed hue since the Pioneer missions, turning from orange to dark brown. A torus of ionized atoms was discovered along Io's orbital path, and volcanoes were found on the moon's surface, some in the process of erupting. As the spacecraft passed behind the planet, it observed flashes of lightning in the night side atmosphere.

The next mission to encounter Jupiter was the *Ulysses* solar probe. It performed a flyby maneuver to attain a polar orbit around the Sun. During this pass, the spacecraft conducted studies on Jupiter's magnetosphere. *Ulysses* has no cameras so no images were taken. A second flyby six years later was at a much greater distance.

In 2000, the *Cassini* probe flew by Jupiter on its way to Saturn, and provided some of the highest-resolution images ever made of the planet.

The *New Horizons* probe flew by Jupiter for a gravity assist en route to Pluto. Its closest approach was on February 28, 2007. The probe's cameras measured plasma output from volcanoes on Io and studied all four Galilean moons in detail, as well as making long-distance observations of the outer moons Himalia and Elara. Imaging of the Jovian system began September 4, 2006.

Galileo mission

The first spacecraft to orbit Jupiter was the *Galileo* probe, which entered orbit on December 7, 1995. It orbited the planet for over seven years, conducting multiple flybys of all the Galilean moons and Amalthea. The spacecraft also witnessed the impact of Comet Shoemaker–Levy 9 as it approached Jupiter in 1994, giving a unique vantage point for the event. Its originally designed capacity was limited by the failed deployment of its high-gain radio antenna, although extensive information was still gained about the Jovian system from *Galileo*.

A 340-kilogram titanium atmospheric probe was released from the spacecraft in July 1995, entering Jupiter's atmosphere on December 7. It parachuted

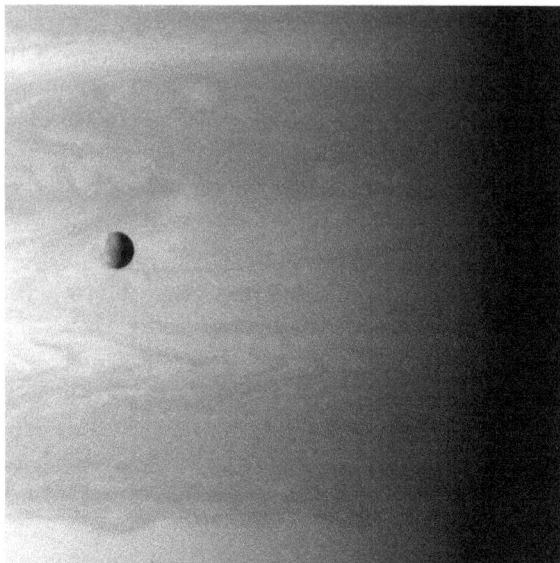

Figure 14: *Cassini views Jupiter and Io on January 1, 2001*

Figure 15: *Jupiter as seen by the space probe Cassini*

through 150 km (93 mi) of the atmosphere at a speed of about 2,575 km/h (1600 mph) and collected data for 57.6 minutes before the signal was lost at a pressure of about 23 atmospheres at a temperature of 153 °C. It melted thereafter, and possibly vaporized. The *Galileo* orbiter itself experienced a more rapid version of the same fate when it was deliberately steered into the planet on September 21, 2003 at a speed of over 50 km/s to avoid any possibility of it crashing into and possibly contaminating Europa, a moon which has been hypothesized to have the possibility of harboring life.

Data from this mission revealed that hydrogen composes up to 90% of Jupiter's atmosphere. The recorded temperature was more than 300 °C (>570 °F) and the windspeed measured more than 644 km/h (>400 mph) before the probes vapourised.

Juno mission

NASA's *Juno* mission arrived at Jupiter on July 4, 2016, and is expected to complete 37 orbits over the next 20 months. The mission plan called for *Juno* to study the planet in detail from a polar orbit. On August 27, 2016, the spacecraft completed its first fly-by of Jupiter and sent back the first-ever images of Jupiter's north pole.

Future probes

The next planned mission to the Jovian system will be the European Space Agency's Jupiter Icy Moon Explorer (JUICE), due to launch in 2022, followed by NASA's Europa Clipper mission in 2025.

Canceled missions

There has been great interest in studying the icy moons in detail because of the possibility of subsurface liquid oceans on Jupiter's moons Europa, Ganymede, and Callisto. Funding difficulties have delayed progress. NASA's *JIMO* (*Jupiter Icy Moons Orbiter*) was cancelled in 2005. A subsequent proposal was developed for a joint NASA/ESA mission called EJSM/Laplace, with a provisional launch date around 2020. EJSM/Laplace would have consisted of the NASA-led Jupiter Europa Orbiter and the ESA-led Jupiter Ganymede Orbiter. However, ESA had formally ended the partnership by April 2011, citing budget issues at NASA and the consequences on the mission timetable. Instead, ESA planned to go ahead with a European-only mission to compete in its L1 Cosmic Vision selection.[13]

Moons

	Wikimedia Commons has media related to *Moons of Jupiter*.

Jupiter has 79 known natural satellites. Of these, 63 are less than 10 kilometres in diameter and have only been discovered since 1975. The four largest moons, visible from Earth with binoculars on a clear night, known as the "Galilean moons", are Io, Europa, Ganymede, and Callisto.

Galilean moons

The moons discovered by Galileo—Io, Europa, Ganymede, and Callisto—are among the largest satellites in the Solar System. The orbits of three of them (Io, Europa, and Ganymede) form a pattern known as a Laplace resonance; for every four orbits that Io makes around Jupiter, Europa makes exactly two orbits and Ganymede makes exactly one. This resonance causes the gravitational effects of the three large moons to distort their orbits into elliptical shapes, because each moon receives an extra tug from its neighbors at the same point in every orbit it makes. The tidal force from Jupiter, on the other hand, works to circularize their orbits.

The eccentricity of their orbits causes regular flexing of the three moons' shapes, with Jupiter's gravity stretching them out as they approach it and allowing them to spring back to more spherical shapes as they swing away. This tidal flexing heats the moons' interiors by friction. This is seen most dramatically in the extraordinary volcanic activity of innermost Io (which is subject to the strongest tidal forces), and to a lesser degree in the geological youth of Europa's surface (indicating recent resurfacing of the moon's exterior).

Name	IPA	Diameter		Mass		Orbital radius		Orbital period	
		km	%	kg	%	km	%	days	%
Io	/ˈaɪ.oʊ/	3,643	105	8.9×10^{22}	120	421,700	110	1.77	7
Europa	/juˈroʊpə/	3,122	90	4.8×10^{22}	65	671,034	175	3.55	13
Ganymede	/ˈgænimiːd/	5,262	150	14.8×10^{22}	200	1,070,412	280	7.15	26
Callisto	/kəˈlɪstoʊ/	4,821	140	10.8×10^{22}	150	1,882,709	490	16.69	61

The Galilean moons Io, Europa, Ganymede, Callisto (in order of increasing distance from Jupiter)

Classification

Before the discoveries of the Voyager missions, Jupiter's moons were arranged neatly into four groups of four, based on commonality of their orbital elements. Since then, the large number of new small outer moons has complicated this picture. There are now thought to be six main groups, although some are more distinct than others.

A basic sub-division is a grouping of the eight inner regular moons, which have nearly circular orbits near the plane of Jupiter's equator and are thought to have formed with Jupiter. The remainder of the moons consist of an unknown number of small irregular moons with elliptical and inclined orbits, which are thought to be captured asteroids or fragments of captured asteroids. Irregular moons that belong to a group share similar orbital elements and thus may have a common origin, perhaps as a larger moon or captured body that broke up.

Regular moons	
Inner group	The inner group of four small moons all have diameters of less than 200 km, orbit at radii less than 200,000 km, and have orbital inclinations of less than half a degree.
Galilean moons	These four moons, discovered by Galileo Galilei and by Simon Marius in parallel, orbit between 400,000 and 2,000,000 km, and are some of the largest moons in the Solar System.
Irregular moons	
Themisto	This is a single moon belonging to a group of its own, orbiting halfway between the Galilean moons and the Himalia group.
Himalia group	A tightly clustered group of moons with orbits around 11,000,000–12,000,000 km from Jupiter.
Carpo	Another isolated case; at the inner edge of the Ananke group, it orbits Jupiter in prograde direction.
S/2016 J 2	A third isolated case, which has a prograde orbit but overlaps the retrograde groups listed below; this may result in a future collision.
Ananke group	This retrograde orbit group has rather indistinct borders, averaging 21,276,000 km from Jupiter with an average inclination of 149 degrees.
Carme group	A fairly distinct retrograde group that averages 23,404,000 km from Jupiter with an average inclination of 165 degrees.
Pasiphae group	A dispersed and only vaguely distinct retrograde group that covers all the outermost moons.

Planetary rings

Jupiter has a faint planetary ring system composed of three main segments: an inner torus of particles known as the halo, a relatively bright main ring, and an outer gossamer ring. These rings appear to be made of dust, rather than ice as with Saturn's rings. The main ring is probably made of material ejected

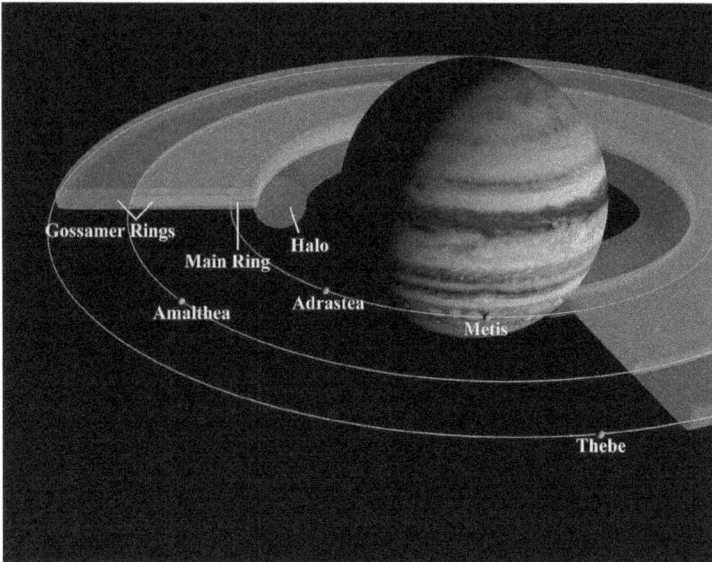

Figure 16: *The rings of Jupiter*

from the satellites Adrastea and Metis. Material that would normally fall back to the moon is pulled into Jupiter because of its strong gravitational influence. The orbit of the material veers towards Jupiter and new material is added by additional impacts. In a similar way, the moons Thebe and Amalthea probably produce the two distinct components of the dusty gossamer ring. There is also evidence of a rocky ring strung along Amalthea's orbit which may consist of collisional debris from that moon.

Interaction with the Solar System

Along with the Sun, the gravitational influence of Jupiter has helped shape the Solar System. The orbits of most of the system's planets lie closer to Jupiter's orbital plane than the Sun's equatorial plane (Mercury is the only planet that is closer to the Sun's equator in orbital tilt), the Kirkwood gaps in the asteroid belt are mostly caused by Jupiter, and the planet may have been responsible for the Late Heavy Bombardment of the inner Solar System's history.

Along with its moons, Jupiter's gravitational field controls numerous asteroids that have settled into the regions of the Lagrangian points preceding and following Jupiter in its orbit around the Sun. These are known as the Trojan asteroids, and are divided into Greek and Trojan "camps" to commemorate the *Iliad*. The first of these, 588 Achilles, was discovered by Max Wolf in

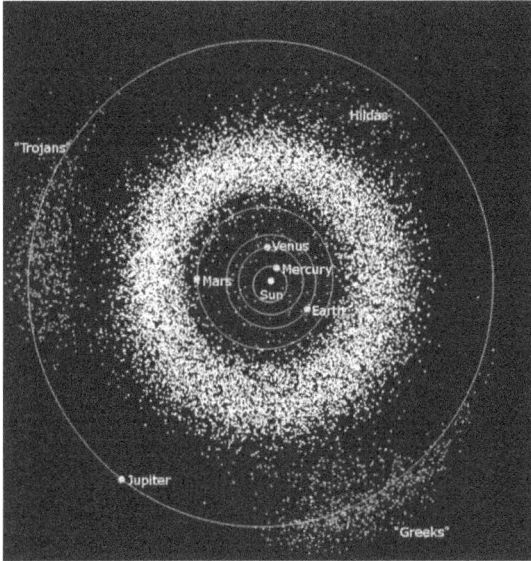

Figure 17: *This diagram shows the Trojan aster-*
oids in Jupiter's orbit, as well as the main asteroid belt.

1906; since then more than two thousand have been discovered. The largest is 624 Hektor.

Most short-period comets belong to the Jupiter family—defined as comets with semi-major axes smaller than Jupiter's. Jupiter family comets are thought to form in the Kuiper belt outside the orbit of Neptune. During close encounters with Jupiter their orbits are perturbed into a smaller period and then circular-ized by regular gravitational interaction with the Sun and Jupiter.

Due to the magnitude of Jupiter's mass, the center of gravity between it and the Sun lies just above the Sun's surface. Jupiter is the only body in the Solar System for which this is true.

Impacts

Jupiter has been called the Solar System's vacuum cleaner, because of its im-mense gravity well and location near the inner Solar System. It receives the most frequent comet impacts of the Solar System's planets. It was thought that the planet served to partially shield the inner system from cometary bom-bardment. However, recent computer simulations suggest that Jupiter does not cause a net decrease in the number of comets that pass through the inner Solar System, as its gravity perturbs their orbits inward roughly as often as it

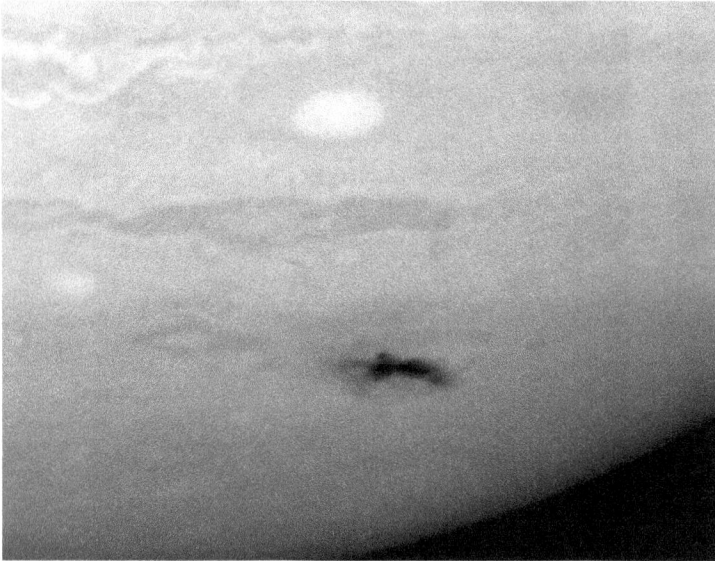

Figure 18: *Hubble image taken on July 23, 2009, showing a blemish of about 8,000 km (5,000 mi) long left by the 2009 Jupiter impact.*

accretes or ejects them. This topic remains controversial among scientists, as some think it draws comets towards Earth from the Kuiper belt while others think that Jupiter protects Earth from the alleged Oort cloud. Jupiter experiences about 200 times more asteroid and comet impacts than Earth.

A 1997 survey of early astronomical records and drawings suggested that a certain dark surface feature discovered by astronomer Giovanni Cassini in 1690 may have been an impact scar. The survey initially produced eight more candidate sites as potential impact observations that he and others had recorded between 1664 and 1839. It was later determined, however, that these candidate sites had little or no possibility of being the results of the proposed impacts.

More recent discoveries include the following:

1. A fireball was photographed by *Voyager 1* during its Jupiter encounter in March 1979.
2. During the period July 16, 1994, to July 22, 1994, over 20 fragments from the comet Shoemaker–Levy 9 (SL9, formally designated D/1993 F2) collided with Jupiter's southern hemisphere, providing the first direct observation of a collision between two Solar System objects. This impact provided useful data on the composition of Jupiter's atmosphere.

3. On July 19, 2009, an impact site was discovered at approximately 216 degrees longitude in System 2. This impact left behind a black spot in Jupiter's atmosphere, similar in size to Oval BA. Infrared observation showed a bright spot where the impact took place, meaning the impact warmed up the lower atmosphere in the area near Jupiter's south pole.

4. A fireball, smaller than the previous observed impacts, was detected on June 3, 2010, by Anthony Wesley, an amateur astronomer in Australia, and was later discovered to have been captured on video by another amateur astronomer in the Philippines.

5. Yet another fireball was seen on August 20, 2010.

6. On September 10, 2012, another fireball was detected.

7. On March 17, 2016 an asteroid or comet struck and was filmed on video.

References

Further reading

- Bagenal, F.; Dowling, T. E.; McKinnon, W. B., eds. (2004). *Jupiter: The planet, satellites, and magnetosphere*. Cambridge: Cambridge University Press. ISBN 0-521-81808-7.
- Beebe, Reta (1997). *Jupiter: The Giant Planet* (Second ed.). Washington, D.C.: Smithsonian Institution Press. ISBN 1-56098-731-6.
- Gore, Rick (January 1980). "What Voyager Saw: Jupiter's Dazzling Realm". *National Geographic*. Vol. 157 no. 1. pp. 2–30. ISSN 0027-9358[14]. OCLC 643483454[15].
- Seidelmann, P. Kenneth; Archinal, Brent A.; A'Hearn, Michael F.; et al. (2007). "Report of the IAU/IAG Working Group on cartographic coordinates and rotational elements: 2006"[16]. *Celestial Mechanics and Dynamical Astronomy*. **98** (3): 155–180. Bibcode: 2007CeMDA.. 98..155S[17]. doi: 10.1007/s10569-007-9072-y[18].
- de Pater, Imke; Lissauer, Jack J. (2015). *Planetary Sciences*[19] (2nd updated ed.). New York: Cambridge University Press. p. 250. ISBN 978-0521853712.

External links

- Hans Lohninger; et al. (November 2, 2005). "Jupiter, As Seen By Voyager 1"[20]. *A Trip into Space*[21]. Virtual Institute of Applied Science. Retrieved March 9, 2007.
- Dunn, Tony (2006). "The Jovian System"[22]. *Gravity Simulator*. Retrieved March 9, 2007. — A simulation of the 62 moons of Jupiter.

- Seronik, G.; Ashford, A. R. "Chasing the Moons of Jupiter"[23]. *Sky & Telescope*. Archived from the original[24] on December 10, 2012. Retrieved March 9, 2007.
- "In Pictures: New views of Jupiter"[25]. BBC News. May 2, 2007. Retrieved May 2, 2007.
- Cain, Fraser. "Jupiter"[26]. *Universe Today*. Retrieved April 1, 2008.
- "Fantastic Flyby of the New Horizons spacecraft (May 1, 2007.)"[27]. NASA. Archived from the original[28] on October 20, 2011. Retrieved May 21, 2008.
- "Moons of Jupiter articles in Planetary Science Research Discoveries"[29]. *Planetary Science Research Discoveries*. University of Hawaii, NASA. Retrieved November 17, 2015.
- Jupiter in Motion[30], album of *Juno* imagery stitched into short videos
- June 2010 impact video[31]
- Bauer, Amanda; Merrifield, Michael (2009). "Jupiter"[32]. *Sixty Symbols*. Brady Haran for the University of Nottingham.
- "NASA Solar System Jupiter"[33].
- Photographs of Jupiter circa 1920s from the Lick Observatory Records Digital Archive, UC Santa Cruz Library's Digital Collections[34]

Mass and size

Jupiter mass

Jovian Mass	
Estimated relative size of the planet Jupiter and the brown dwarfs WISE-1828, Gliese 229B, and Teide 1 compared to the Sun and a red dwarf star.	

Unit information	
Unit system	Astronomical system of units
Unit of	mass
Symbol	M_J or M_{Jup}, $M_{2\!\!\!+}$

Unit conversions	
1 M_J in ...	*... is equal to ...*
SI base unit	$(1.89813 \pm 0.00019) \times 10^{27}$ kg
U.S. customary	$\approx 4.1847 \times 10^{27}$ pounds

Jupiter mass, also called **Jovian mass** is the unit of mass equal to the total mass of the planet Jupiter. This value may refer to the mass of planet Jupiter alone, or the mass of the entire Jovian system to include the Moons of Jupiter. Jupiter is by far the most massive planet in the solar system. It is approximately 2.5 times more massive than all of the other planets in the Solar System combined.

Jupiter mass is a common unit of mass in astronomy that is used to indicate the masses of other similarly-sized objects, including the outer planets and extrasolar planets. It may also be used to describe the masses of brown dwarfs, as this unit provides a convenient scale for comparison.

Current best estimates

The current best known value for the mass of Jupiter can be expressed as 1898130 yottagrams:

$$M_J = (1.89813 \pm 0.00019) \times 10^{27} \text{ kg,}$$

which is about 1000 times less massive than the sun (about 0.1% $M_\odot$):

$$M_J = \frac{1}{1047.348644 \pm 0.000017} M_{Sun}$$

in scientific notation,

$$M_J = 954.7919(15) \times 10^{-6} M_{Sun}.$$

Compared to the mass of Earth, Jupiter is 318 times more massive.

$$M_J = 317.82838 \ M_\oplus.$$

Context and implications

Jupiter's mass is 2.5 times that of all the other planets in the Solar System combined—this is so massive that its barycenter with the Sun lies beyond the Sun's surface at 1.068 solar radii from the Sun's center.

Because the mass of Jupiter is so large compared to the other objects in the solar system, the effects of its gravity must be included when calculating satellite trajectories and the precise orbits of other bodies in the solar system, including Earth's moon and even Pluto.

Theoretical models indicate that if Jupiter had much more mass than it does at present, its atmosphere would collapse, and the planet would shrink. For small changes in mass, the radius would not change appreciably, but above about 500 $M_\oplus$ (1.6 Jupiter masses) the interior would become so much more compressed under the increased pressure that its volume would *decrease* despite the increasing amount of matter. As a result, Jupiter is thought to have about as large a diameter as a planet of its composition and evolutionary history can achieve. The process of further shrinkage with increasing mass would continue until appreciable stellar ignition was achieved, as in high-mass brown dwarfs having around 50 Jupiter masses. Jupiter would need to be about 75 times more massive to fuse hydrogen and become a star.

Gravitational constant

The mass of Jupiter is derived from the measured value called the Jovian mass parameter, which is denoted with GM_J. The mass of Jupiter is calculated by dividing GM_J by the constant G. For celestial bodies such as Jupiter, Earth and the Sun, the value of the GM product is known to many orders of magnitude more precisely than either factor independently. The limited precision available for G limits the uncertainty of the derived mass. For this reason, astronomers often prefer to refer to the gravitational parameter, rather than the explicit mass. The GM products are used when computing the ratio of Jupiter mass relative to other objects.

In 2015, the International Astronomical Union defined the *nominal Jovian mass parameter* to remain constant regardless of subsequent improvements in measurement precision of M_J. This constant is defined as exactly:

$$(\mathcal{GM})_J^N = 1.2668653 \times 10^{17} \text{ m}^3/\text{s}^2$$

If the explicit mass of Jupiter is needed in SI units, it can be calculated in terms of the Gravitational constant, G by dividing GM by G.

Mass composition

The majority of Jupiter's mass is hydrogen and helium. These two elements make up more than 87% of the total mass of Jupiter. The total mass of heavy elements other than hydrogen and helium in the planet is between 11 and 45 $M_\oplus$. The bulk of the hydrogen on Jupiter is solid hydrogen. Evidence suggests that Jupiter contains a central dense core. If so, the mass of the core is predicted to be no larger than about 12 $M_\oplus$. The exact mass of the core is uncertain due to the relatively poor knowledge of the behavior of solid hydrogen at very high pressures.

Relative mass

Masses of noteworthy astronomical objects relative to the mass of Jupiter

Object	M_J / M_{object}	M_{object} / M_J	Ref
Sun	$954.7919(15) \times 10^{-6}$	1047.348644 ± 0.000017	
Earth	317.82838	0.0031463520	
Jupiter	1	1	by definition
Saturn	3.3397683	0.29942197	[35]
Uranus	21.867552	0.045729856	

Neptune	18.53467	0.05395295	
Gliese 229B		21–52.4	
51 Pegasi b		0.472±0.039	

Atmosphere

Atmosphere of Jupiter

The **atmosphere of Jupiter** is the largest planetary atmosphere in the Solar System. It is mostly made of molecular hydrogen and helium in roughly solar proportions; other chemical compounds are present only in small amounts and include methane, ammonia, hydrogen sulfide and water. Although water is thought to reside deep in the atmosphere, its directly measured concentration is very low. The nitrogen, sulfur, and noble gas abundances in Jupiter's atmosphere exceed solar values by a factor of about three.[36]

The atmosphere of Jupiter lacks a clear lower boundary and gradually transitions into the liquid interior of the planet. From lowest to highest, the atmospheric layers are the troposphere, stratosphere, thermosphere and exosphere. Each layer has characteristic temperature gradients. The lowest layer, the troposphere, has a complicated system of clouds and hazes, comprising layers of ammonia, ammonium hydrosulfide and water.[37] The upper ammonia clouds visible at Jupiter's surface are organized in a dozen zonal bands parallel to the equator and are bounded by powerful zonal atmospheric flows (winds) known as *jets*. The bands alternate in color: the dark bands are called *belts*, while light ones are called *zones*. Zones, which are colder than belts, correspond to upwellings, while belts mark descending gas. The zones' lighter color is believed to result from ammonia ice; what gives the belts their darker colors is uncertain. The origins of the banded structure and jets are not well understood, though a "shallow model" and a "deep model" exist.

The Jovian atmosphere shows a wide range of active phenomena, including band instabilities, vortices (cyclones and anticyclones), storms and lightning. The vortices reveal themselves as large red, white or brown spots (ovals). The largest two spots are the Great Red Spot (GRS) and Oval BA, which is also red. These two and most of the other large spots are anticyclonic. Smaller anticyclones tend to be white. Vortices are thought to be relatively shallow

Figure 19: *Jupiter's swirling colourful clouds*

structures with depths not exceeding several hundred kilometers. Located in the southern hemisphere, the GRS is the largest known vortex in the Solar System. It could engulf two or three Earths and has existed for at least three hundred years. Oval BA, south of GRS, is a red spot a third the size of GRS that formed in 2000 from the merging of three white ovals.

Jupiter has powerful storms, often accompanied by lightning strikes. The storms are a result of moist convection in the atmosphere connected to the evaporation and condensation of water. They are sites of strong upward motion of the air, which leads to the formation of bright and dense clouds. The storms form mainly in belt regions. The lightning strikes on Jupiter are hundreds of times more powerful than those seen on Earth, and are assumed to be associated with the water clouds. This storm near the red spot is called Red Spot Junior.

Vertical structure

The atmosphere of Jupiter is classified into four layers, by increasing altitude: the troposphere, stratosphere, thermosphere and exosphere. Unlike the Earth's atmosphere, Jupiter's lacks a mesosphere. Jupiter does not have a solid surface, and the lowest atmospheric layer, the troposphere, smoothly transitions into the planet's fluid interior. This is a result of having temperatures and the pressures

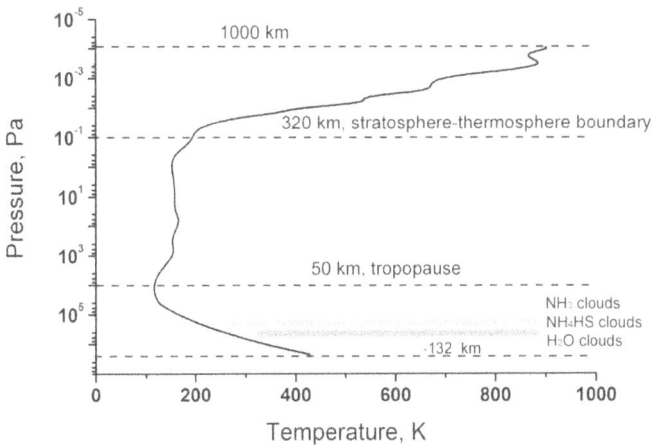

Figure 20: *Vertical structure of the atmosphere of Jupiter. Note that the temperature drops together with altitude above the tropopause. The Galileo atmospheric probe stopped transmitting at a depth of 132 km below the 1 bar "surface" of Jupiter.*

well above those of the critical points for hydrogen and helium, meaning that there is no sharp boundary between gas and liquid phases. Hydrogen becomes a supercritical fluid at a pressure of around 12 bar.[38]

Since the lower boundary of the atmosphere is ill-defined, the pressure level of 10 bars, at an altitude of about 90 km below 1 bar with a temperature of around 340 K, is commonly treated as the base of the troposphere. In scientific literature, the 1 bar pressure level is usually chosen as a zero point for altitudes—a "surface" of Jupiter. As with Earth, the top atmospheric layer, the exosphere, does not have a well defined upper boundary.[39] The density gradually decreases until it smoothly transitions into the interplanetary medium approximately 5,000 km above the "surface".[40]

The vertical temperature gradients in the Jovian atmosphere are similar to those of the atmosphere of Earth. The temperature of the troposphere decreases with height until it reaches a minimum at the tropopause, which is the boundary between the troposphere and stratosphere. On Jupiter, the tropopause is approximately 50 km above the visible clouds (or 1 bar level), where the pressure and temperature are about 0.1 bar and 110 K. In the stratosphere, the temperatures rise to about 200 K at the transition into the thermosphere, at an altitude and pressure of around 320 km and 1 μbar.[41] In the thermosphere, temperatures continue to rise, eventually reaching 1000 K at about 1000 km, where pressure is about 1 nbar.

Jupiter's troposphere contains a complicated cloud structure.[42] The upper clouds, located in the pressure range 0.6–0.9 bar, are made of ammonia ice.[43] Below these ammonia ice clouds, denser clouds made of ammonium hydrosulfide ($(NH_4)SH$) or ammonium sulfide ($(NH_4)_2S$, between 1–2 bar) and water (3–7 bar) are thought to exist.[44] There are no methane clouds as the temperatures are too high for it to condense.[42] The water clouds form the densest layer of clouds and have the strongest influence on the dynamics of the atmosphere. This is a result of the higher condensation heat of water and higher water abundance as compared to the ammonia and hydrogen sulfide (oxygen is a more abundant chemical element than either nitrogen or sulfur).[45] Various tropospheric (at 200–500 mbar) and stratospheric (at 10–100 mbar) haze layers reside above the main cloud layers.[46] The latter are made from condensed heavy polycyclic aromatic hydrocarbons or hydrazine, which are generated in the upper stratosphere (1–100 µbar) from methane under the influence of the solar ultraviolet radiation (UV).[42] The methane abundance relative to molecular hydrogen in the stratosphere is about 10^{-4},[40] while the abundance ratio of other light hydrocarbons, like ethane and acetylene, to molecular hydrogen is about 10^{-6}.[40]

Jupiter's thermosphere is located at pressures lower than 1 µbar and demonstrates such phenomena as airglow, polar aurorae and X-ray emissions.[47] Within it lie layers of increased electron and ion density that form the ionosphere.[40] The high temperatures prevalent in the thermosphere (800–1000 K) have not been fully explained yet;[48] existing models predict a temperature no higher than about 400 K.[40] They may be caused by absorption of high-energy solar radiation (UV or X-ray), by heating from the charged particles precipitating from the Jovian magnetosphere, or by dissipation of upward-propagating gravity waves.[49] The thermosphere and exosphere at the poles and at low latitudes emit X-rays, which were first observed by the Einstein Observatory in 1983.[50] The energetic particles coming from Jupiter's magnetosphere create bright auroral ovals, which encircle the poles. Unlike their terrestrial analogs, which appear only during magnetic storms, aurorae are permanent features of Jupiter's atmosphere.[50] The thermosphere was the first place outside the Earth where the trihydrogen cation (H^+_3) was discovered.[40] This ion emits strongly in the mid-infrared part of the spectrum, at wavelengths between 3 and 5 µm; this is the main cooling mechanism of the thermosphere.

Chemical composition

Elemental abundances relative to hydrogen
in Jupiter and Sun[36]

Element	Sun	Jupiter/Sun
He/H	0.0975	0.807 ± 0.02
Ne/H	1.23×10^{-4}	0.10 ± 0.01
Ar/H	3.62×10^{-6}	2.5 ± 0.5
Kr/H	1.61×10^{-9}	2.7 ± 0.5
Xe/H	1.68×10^{-10}	2.6 ± 0.5
C/H	3.62×10^{-4}	2.9 ± 0.5
N/H	1.12×10^{-4}	3.6 ± 0.5 (8 bar) 3.2 ± 1.4 (9–12 bar)
O/H	8.51×10^{-4}	0.033 ± 0.015 (12 bar) 0.19–0.58 (19 bar)
P/H	3.73×10^{-7}	0.82
S/H	1.62×10^{-5}	2.5 ± 0.15

Isotopic ratios in Jupiter and Sun[36]

Ratio	Sun	Jupiter
$^{13}C/^{12}C$	0.011	0.0108 ± 0.0005
$^{15}N/^{14}N$	$<2.8 \times 10^{-3}$	$2.3 \pm 0.3 \times 10^{-3}$ (0.08–2.8 bar)
$^{36}Ar/^{38}Ar$	5.77 ± 0.08	5.6 ± 0.25
$^{20}Ne/^{22}Ne$	13.81 ± 0.08	13 ± 2
$^{3}He/^{4}He$	$1.5 \pm 0.3 \times 10^{-4}$	$1.66 \pm 0.05 \times 10^{-4}$
D/H	$3.0 \pm 0.17 \times 10^{-5}$	$2.25 \pm 0.35 \times 10^{-5}$

The composition of Jupiter's atmosphere is similar to that of the planet as a whole.[36] Jupiter's atmosphere is the most comprehensively understood of those of all the gas giants because it was observed directly by the *Galileo* atmospheric probe when it entered the Jovian atmosphere on December 7, 1995. Other sources of information about Jupiter's atmospheric composition include the Infrared Space Observatory (ISO),[51] the *Galileo* and *Cassini* orbiters,[52] and Earth-based observations.[36]

The two main constituents of the Jovian atmosphere are molecular hydrogen (H$_2$) and helium.[36] The helium abundance is 0.157 ± 0.004 relative to molecular hydrogen by number of molecules, and its mass fraction is 0.234 ± 0.005,

which is slightly lower than the Solar System's primordial value.[36] The reason
for this low abundance is not entirely understood, but some of the helium may
have condensed into the core of Jupiter.[43] This condensation is likely to be
in the form of helium rain: as hydrogen turns into the metallic state at depths
of more than 10,000 km, helium separates from it forming droplets which,
being denser than the metallic hydrogen, descend towards the core. This can
also explain the severe depletion of neon (see Table), an element that easily
dissolves in helium droplets and would be transported in them towards the
core as well.

The atmosphere contains various simple compounds such as water, methane
(CH_4), hydrogen sulfide (H_2S), ammonia (NH_3) and phosphine (PH_3).[36]
Their abundances in the deep (below 10 bar) troposphere imply that the atmosphere of Jupiter is enriched in the elements carbon, nitrogen, sulfur and
possibly oxygen[b] by factor of 2–4 relative to the Sun.[c][36] The noble gases
argon, krypton and xenon also appear in abundance relative to solar levels
(see table), while neon is scarcer.[36] Other chemical compounds such as ar-
sine (AsH_3) and germane (GeH_4) are present only in trace amounts.[36] The
upper atmosphere of Jupiter contains small amounts of simple hydrocarbons
such as ethane, acetylene, and diacetylene, which form from methane under
the influence of the solar ultraviolet radiation and charged particles coming
from Jupiter's magnetosphere.[36] The carbon dioxide, carbon monoxide and
water present in the upper atmosphere are thought to originate from impacting
comets, such as Shoemaker-Levy 9. The water cannot come from the troposphere because the cold tropopause acts like a cold trap, effectively preventing
water from rising to the stratosphere (see Vertical structure above).[36]

Earth- and spacecraft-based measurements have led to improved knowledge
of the isotopic ratios in Jupiter's atmosphere. As of July 2003, the accepted
value for the deuterium abundance is $(2.25 \pm 0.35) \times 10^{-5}$,[36] which probably represents the primordial value in the protosolar nebula that gave birth to
the Solar System.[51] The ratio of nitrogen isotopes in the Jovian atmosphere,
^{15}N to ^{14}N, is 2.3×10^{-3}, a third lower than that in the Earth's atmosphere
(3.5×10^{-3}).[36] The latter discovery is especially significant since the previous theories of Solar System formation considered the terrestrial value for the
ratio of nitrogen isotopes to be primordial.[51]

Zones, belts and jets

The visible surface of Jupiter is divided into several bands parallel to the equator. There are two types of bands: lightly colored *zones* and relatively dark
belts.[53] The wider Equatorial Zone (EZ) extends between latitudes of approximately 7°S to 7°N. Above and below the EZ, the North and South Equatorial belts (NEB and SEB) extend to 18°N and 18°S, respectively. Farther

Figure 21: *Polar view of planet Jupiter*

from the equator lie the North and South Tropical zones (NtrZ and STrZ). The alternating pattern of belts and zones continues until the polar regions at approximately 50 degrees latitude, where their visible appearance becomes somewhat muted.[54] The basic belt-zone structure probably extends well towards the poles, reaching at least to 80° North or South.

The difference in the appearance between zones and belts is caused by differences in the opacity of the clouds. Ammonia concentration is higher in zones, which leads to the appearance of denser clouds of ammonia ice at higher altitudes, which in turn leads to their lighter color. On the other hand, in belts clouds are thinner and are located at lower altitudes.[55] The upper troposphere is colder in zones and warmer in belts. The exact nature of chemicals that make Jovian zones and bands so colorful is not known, but they may include complicated compounds of sulfur, phosphorus and carbon.

The Jovian bands are bounded by zonal atmospheric flows (winds), called *jets*. The eastward (prograde) jets are found at the transition from zones to belts (going away from the equator), whereas westward (retrograde) jets mark the transition from belts to zones. Such flow velocity patterns mean that the zonal winds decrease in belts and increase in zones from the equator to the pole. Therefore, wind shear in belts is cyclonic, while in zones it is anticyclonic.[56] The EZ is an exception to this rule, showing a strong eastward (prograde) jet

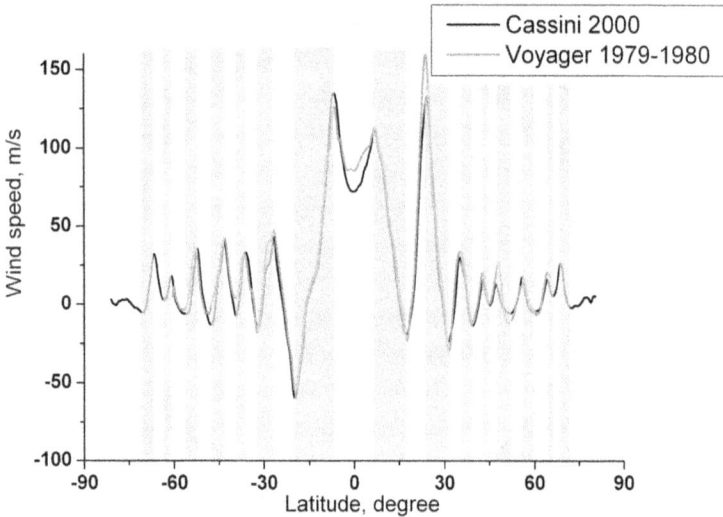

Figure 22: *Zonal wind speeds in the atmosphere of Jupiter*

and has a local minimum of the wind speed exactly at the equator. The jet speeds are high on Jupiter, reaching more than 100 m/s. These speeds correspond to ammonia clouds located in the pressure range 0.7–1 bar. The prograde jets are generally more powerful than the retrograde jets. The vertical extent of jets is not known. They decay over two to three scale heights[a] above the clouds, while below the cloud level, winds increase slightly and then remain constant down to at least 22 bar—the maximum operational depth reached by the Galileo Probe.[57]

The origin of Jupiter's banded structure is not completely clear, though it may be similar to that driving the Earth's Hadley cells. The simplest interpretation is that zones are sites of atmospheric upwelling, whereas belts are manifestations of downwelling. When air enriched in ammonia rises in zones, it expands and cools, forming high and dense clouds. In belts, however, the air descends, warming adiabatically as in a convergence zone on Earth, and white ammonia clouds evaporate, revealing lower, darker clouds. The location and width of bands, speed and location of jets on Jupiter are remarkably stable, having changed only slightly between 1980 and 2000. One example of change is a decrease of the speed of the strongest eastward jet located at the boundary between the North Tropical zone and North Temperate belts at 23°N.[58] However bands vary in coloration and intensity over time (see below). These variations were first observed in the early seventeenth century.[59]

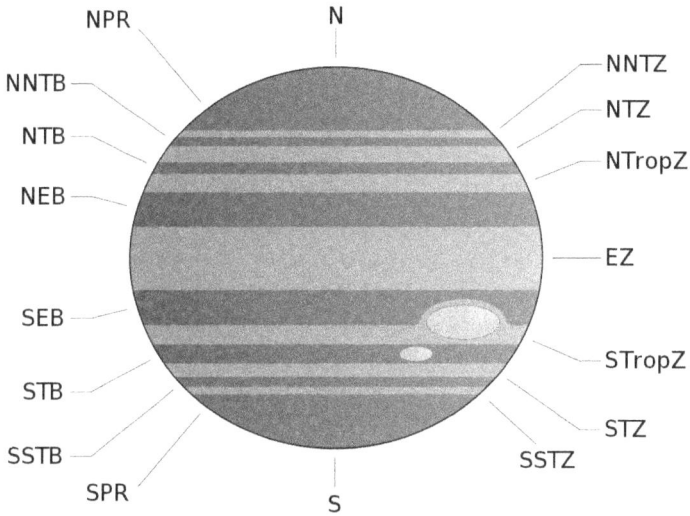

Figure 23: *Idealized illustration of Jupiter's cloud bands, labeled with their official abbreviations. Lighter zones are indicated to the right, darker belts to the left. The Great Red Spot and Oval BA are shown in the South Tropical Zone and South Temperate Belt, respectively.*

Specific bands

The belts and zones that divide Jupiter's atmosphere each have their own names and unique characteristics. They begin below the North and South Polar Regions, which extend from the poles to roughly 40–48° N/S. These bluish-gray regions are usually featureless.

The North North Temperate Region rarely shows more detail than the polar regions, due to limb darkening, foreshortening, and the general diffuseness of features. However, the North-North Temperate Belt (NNTB) is the northern-most distinct belt, though it occasionally disappears. Disturbances tend to be minor and short-lived. The North-North Temperate Zone (NNTZ) is perhaps more prominent, but also generally quiet. Other minor belts and zones in the region are occasionally observed.[60]

The North Temperate Region is part of a latitudinal region easily observable from Earth, and thus has a superb record of observation.[61] It also features the strongest prograde jet stream on the planet—a westerly current that forms the southern boundary of the North Temperate Belt (NTB). The NTB fades

Figure 24: *Zones, belts and vortices on Jupiter. The wide equatorial zone is visible in the center surrounded by two dark equatorial belts (SEB and NEB). The large grayish-blue irregular "hot spots" at the northern edge of the white Equatorial Zone change over the course of time as they march east-ward across the planet. The Great Red Spot is at the southern margin of the SEB. Strings of small storms rotate around northern-hemisphere ovals. Small, very bright features, possible lightning storms, appear quickly and randomly in turbulent regions. The smallest features visible at the equator are about 600 kilometers across. This 14-frame animation spans 24 Jovian days, or about 10 Earth days. The passage of time is accelerated by a factor of 600,000. The occasional black spots in the image are shadows cast by moons.*

roughly once a decade (this was the case during the *Voyager* encounters), making the North Temperate Zone (NTZ) apparently merge into the North Tropical Zone (NTropZ). Other times, the NTZ is divided by a narrow belt into northern and southern components.

The North Tropical Region is composed of the NTropZ and the North Equatorial Belt (NEB). The NTropZ is generally stable in coloration, changing in tint only in tandem with activity on the NTB's southern jet stream. Like the NTZ, it too is sometimes divided by a narrow band, the NTropB. On rare occasions, the southern NTropZ plays host to "Little Red Spots". As the name suggests, these are northern equivalents of the Great Red Spot. Unlike the GRS, they tend to occur in pairs and are always short-lived, lasting a year on average; one was present during the *Pioneer 10* encounter.[62]

The NEB is one of the most active belts on the planet. It is characterized by anticyclonic white ovals and cyclonic "barges" (also known as "brown ovals"), with the former usually forming farther north than the latter; as in the NTropZ, most of these features are relatively short-lived. Like the South Equatorial Belt (SEB), the NEB has sometimes dramatically faded and "revived". The timescale of these changes is about 25 years.[63]

The Equatorial Region (EZ) is one of the most stable regions of the planet, in latitude and in activity. The northern edge of the EZ hosts spectacular plumes

Figure 25: *This image from the HST reveals a rare wave structure just north of the planet's equator.*

that trail southwest from the NEB, which are bounded by dark, warm (in infrared) features known as festoons (hot spots). Though the southern boundary of the EZ is usually quiescent, observations from the late 19th into the early 20th century show that this pattern was then reversed relative to today. The EZ varies considerably in coloration, from pale to an ochre, or even coppery hue; it is occasionally divided by an Equatorial Band (EB).[64] Features in the EZ move roughly 390 km/h relative to the other latitudes.[65,66]

The South Tropical Region includes the South Equatorial Belt (SEB) and the South Tropical Zone. It is by far the most active region the planet, as it is home to its strongest retrograde jet stream. The SEB is usually the broadest, darkest belt on Jupiter; it is sometimes split by a zone (the SEBZ), and can fade entirely every 3 to 15 years before reappearing in what is known as an SEB Revival cycle. A period of weeks or months following the belt's disappearance, a white spot forms and erupts dark brownish material which is stretched into a new belt by Jupiter's winds. The belt most recently disappeared in May 2010. Another characteristic of the SEB is a long train of cyclonic disturbances following the Great Red Spot. Like the NTropZ, the STropZ is one of the most prominent zones on the planet; not only does it contain the GRS, but it is occasionally rent by a South Tropical Disturbance (STropD), a division of the zone that can be very long-lived; the most famous one lasted from 1901 to 1939.[67]

The South Temperate Region, or South Temperate Belt (STB), is yet another dark, prominent belt, more so than the NTB; until March 2000, its most famous features were the long-lived white ovals BC, DE, and FA, which have since merged to form Oval BA ("Red Jr."). The ovals were part of South Temperate Zone, but they extended into STB partially blocking it. The STB has occasionally faded, apparently due to complex interactions between the white ovals and the GRS. The appearance of the South Temperate Zone (STZ)—the zone in which the white ovals originated—is highly variable.[68]

There are other features on Jupiter that are either temporary or difficult to observe from Earth. The South South Temperate Region is harder to discern even than the NNTR; its detail is subtle and can only be studied well by large telescopes or spacecraft.[69] Many zones and belts are more transient in nature and are not always visible. These include the *Equatorial band* (EB),[70] *North Equatorial belt zone* (NEBZ, a white zone within the belt) and *South Equatorial belt zone* (SEBZ).[71] Belts are also occasionally split by a sudden disturbance. When a disturbance divides a normally singular belt or zone, an *N* or an *S* is added to indicate whether the component is the northern or southern one; e.g., NEB(N) and NEB(S).[72]

Dynamics

<templatestyles src="Multiple_image/styles.css" />

2009

2010

Circulation in Jupiter's atmosphere is markedly different from that in the atmosphere of Earth. The interior of Jupiter is fluid and lacks any solid surface. Therefore, convection may occur throughout the planet's outer molecular envelope. As of 2008, a comprehensive theory of the dynamics of the Jovian atmosphere has not been developed. Any such theory needs to explain the following facts: the existence of narrow stable bands and jets that are symmetric relative to Jupiter's equator, the strong prograde jet observed at the equator, the difference between zones and belts, and the origin and persistence of large vortices such as the Great Red Spot.[73]

The theories regarding the dynamics of the Jovian atmosphere can be broadly divided into two classes: shallow and deep. The former hold that the observed circulation is largely confined to a thin outer (weather) layer of the planet, which overlays the stable interior. The latter hypothesis postulates that the observed atmospheric flows are only a surface manifestation of deeply rooted circulation in the outer molecular envelope of Jupiter.[74] As both theories have their own successes and failures, many planetary scientists think that the true theory will include elements of both models.[75]

Shallow models

The first attempts to explain Jovian atmospheric dynamics date back to the 1960s.[76] They were partly based on terrestrial meteorology, which had become well developed by that time. Those shallow models assumed that the jets on Jupiter are driven by small scale turbulence, which is in turn maintained by moist convection in the outer layer of the atmosphere (above the water clouds).[77] The moist convection is a phenomenon related to the condensation and evaporation of water and is one of the major drivers of terrestrial weather.[78] The production of the jets in this model is related to a well-known property of two dimensional turbulence—the so-called inverse cascade, in which small turbulent structures (vortices) merge to form larger ones.[79] The finite size of the planet means that the cascade can not produce structures larger than some characteristic scale, which for Jupiter is called the Rhines scale. Its existence is connected to production of Rossby waves. This process works as follows: when the largest turbulent structures reach a certain size, the energy begins to flow into Rossby waves instead of larger structures, and the inverse cascade stops.[80] Since on the spherical rapidly rotating planet the dispersion relation of the Rossby waves is anisotropic, the Rhines scale in the direction parallel to the equator is larger than in the direction orthogonal to it. The ultimate result of the process described above is production of large scale elongated structures, which are parallel to the equator. The meridional extent of them appears to match the actual width of jets. Therefore, in shallow models vortices actually feed the jets and should disappear by merging into them.

While these weather–layer models can successfully explain the existence of a dozen narrow jets, they have serious problems. A glaring failure of the model is the prograde (super-rotating) equatorial jet: with some rare exceptions shallow models produce a strong retrograde (subrotating) jet, contrary to observations.

Figure 26: *Thermal image of Jupiter obtained by NASA Infrared Telescope Facility*

In addition, the jets tend to be unstable and can disappear over time. Shallow models cannot explain how the observed atmospheric flows on Jupiter violate stability criteria.[81] More elaborated multilayer versions of weather–layer models produce more stable circulation, but many problems persist.[82] Meanwhile, the Galileo Probe found that the winds on Jupiter extend well below the water clouds at 5–7 bar and do not show any evidence of decay down to 22 bar pressure level, which implies that circulation in the Jovian atmosphere may in fact be deep.

Deep models

The deep model was first proposed by Busse in 1976.[83,84] His model was based on another well-known feature of fluid mechanics, the Taylor–Proudman theorem. It holds that in any fast-rotating barotropic ideal liquid, the flows are organized in a series of cylinders parallel to the rotational axis. The conditions of the theorem are probably met in the fluid Jovian interior. Therefore, the planet's molecular hydrogen mantle may be divided into cylinders, each cylinder having a circulation independent of the others.[85] Those latitudes where the cylinders' outer and inner boundaries intersect with the visible surface of the planet correspond to the jets; the cylinders themselves are observed as zones and belts.

The deep model easily explains the strong prograde jet observed at the equator of Jupiter; the jets it produces are stable and do not obey the 2D stability criterion. However it has major difficulties; it produces a very small number of broad jets, and realistic simulations of 3D flows are not possible as of 2008, meaning that the simplified models used to justify deep circulation may fail to catch important aspects of the fluid dynamics within Jupiter. One model published in 2004 successfully reproduced the Jovian band-jet structure. It assumed that the molecular hydrogen mantle is thinner than in all other models; occupying only the outer 10% of Jupiter's radius. In standard models of the Jovian interior, the mantle comprises the outer 20–30%.[86] The driving of deep circulation is another problem. The deep flows can be caused both by shallow forces (moist convection, for instance) or by deep planet-wide convection that transports heat out of the Jovian interior. Which of these mechanisms is more important is not clear yet.

Internal heat

As has been known since 1966,[87] Jupiter radiates much more heat than it receives from the Sun. It is estimated that the ratio between the power emitted by the planet and that absorbed from the Sun is 1.67 ± 0.09. The internal heat flux from Jupiter is 5.44 ± 0.43 W/m^2, whereas the total emitted power is 335 ± 26 petawatts. The latter value is approximately equal to one billionth of the total power radiated by the Sun. This excess heat is mainly the primordial heat from the early phases of Jupiter's formation, but may result in part from the precipitation of helium into the core.[88]

The internal heat may be important for the dynamics of the Jovian atmosphere. While Jupiter has a small obliquity of about $3°$, and its poles receive much less solar radiation than its equator, the tropospheric temperatures do not change appreciably from the equator to poles. One explanation is that Jupiter's convective interior acts like a thermostat, releasing more heat near the poles than in the equatorial region. This leads to a uniform temperature in the troposphere. While heat is transported from the equator to the poles mainly via the atmosphere on Earth, on Jupiter deep convection equilibrates heat. The convection in the Jovian interior is thought to be driven mainly by the internal heat.[89]

Figure 27: *New Horizons IR view of Jupiter's atmosphere*

Discrete features

Vortices

The atmosphere of Jupiter is home to hundreds of vortices—circular rotating structures that, as in the Earth's atmosphere, can be divided into two classes: cyclones and anticyclones.[90] Cyclones rotate in the direction similar to the rotation of the planet (counterclockwise in the northern hemisphere and clockwise in the southern); anticyclones rotate in the reverse direction. However, unlike in the terrestrial atmosphere, anticyclones predominate over cyclones on Jupiter—more than 90% of vortices larger than 2000 km in diameter are anticyclones.[91] The lifetime of Jovian vortices varies from several days to hundreds of years, depending on their size. For instance, the average lifetime of an anticyclone between 1000 and 6000 km in diameter is 1–3 years.[92] Vortices have never been observed in the equatorial region of Jupiter (within 10° of latitude), where they are unstable.[93] As on any rapidly rotating planet, Jupiter's anticyclones are high pressure centers, while cyclones are low pressure.[94]

The anticyclones in Jupiter's atmosphere are always confined within zones, where the wind speed increases in direction from the equator to the poles. They are usually bright and appear as white ovals. They can move in longitude, but stay at approximately the same latitude as they are unable to escape from

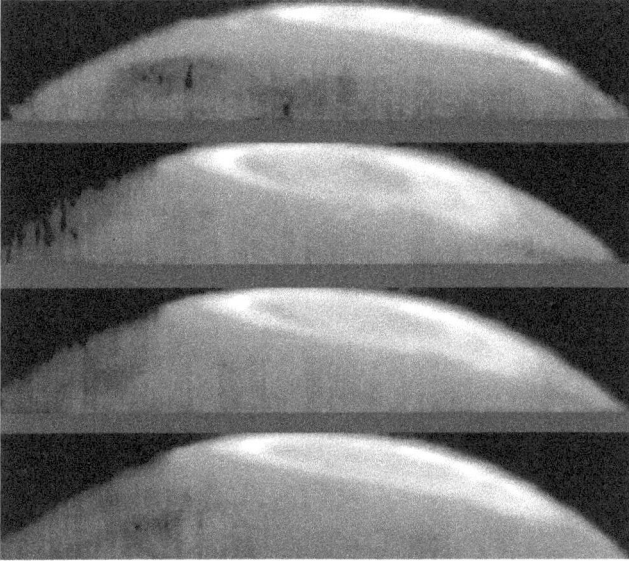

Figure 28: *Great Cold Spot on Jupiter*

Figure 29: *Jupiter clouds*
(Juno; October 2017)

the confining zone. The wind speeds at their periphery are about 100 m/s. Different anticyclones located in one zone tend to merge, when they approach each other.[95] However Jupiter has two anticyclones that are somewhat different from all others. They are the Great Red Spot (GRS)[96] and the Oval BA;[97] the latter formed only in 2000. In contrast to white ovals, these structures are red, arguably due to dredging up of red material from the planet's depths. On Jupiter the anticyclones usually form through merges of smaller structures including convective storms (see below), although large ovals can result from the instability of jets. The latter was observed in 1938–1940, when a few white ovals appeared as a result of instability of the southern temperate zone; they later merged to form Oval BA.

In contrast to anticyclones, the Jovian cyclones tend to be small, dark and irregular structures. Some of the darker and more regular features are known as brown ovals (or badges). However the existence of a few long–lived large cyclones has been suggested. In addition to compact cyclones, Jupiter has several large irregular filamentary patches, which demonstrate cyclonic rotation. One of them is located to the west of the GRS (in its wake region) in the southern equatorial belt.[98] These patches are called cyclonic regions (CR). The cyclones are always located in the belts and tend to merge when they encounter each other, much like anticyclones.

The deep structure of vortices is not completely clear. They are thought to be relatively thin, as any thickness greater than about 500 km will lead to instability. The large anticyclones are known to extend only a few tens of kilometers above the visible clouds. The early hypothesis that the vortices are deep convective plumes (or convective columns) as of 2008 is not shared by the majority of planetary scientists.

Great Red Spot

The Great Red Spot (GRS) is a persistent anticyclonic storm, 22° south of Jupiter's equator; observations from Earth establish a minimum storm lifetime of 350 years. A storm was described as a "permanent spot" by Gian Domenico Cassini after observing the feature in July 1665 with his instrument-maker Eustachio Divini. According to a report by Giovanni Battista Riccioli in 1635, Leander Bandtius, whom Riccioli identified as the Abbot of Dunisburgh who possessed an "extraordinary telescope", observed a large spot that he described as "oval, equaling one seventh of Jupiter's diameter at its longest." According to Riccioli, "these features are seldom able to be seen, and then only by a telescope of exceptional quality and magnification."[99] The Great Spot has been nearly continually observed since the 1870s, however.

The GRS rotates counter-clockwise, with a period of about six Earth days[100] or 14 Jovian days. Its dimensions are 24,000–40,000 km east-to-west and

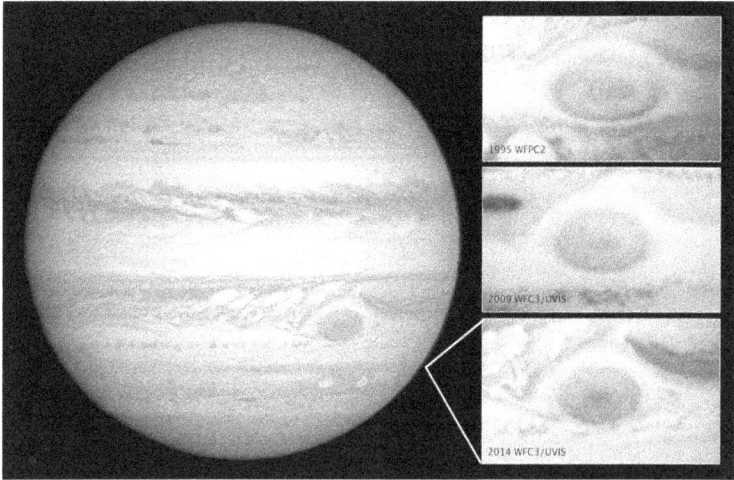

Figure 30: *The Great Red Spot is decreasing in size (May 15, 2014).*

12,000–14,000 km north-to-south. The spot is large enough to contain two or three planets the size of Earth. At the start of 2004, the Great Red Spot had approximately half the longitudinal extent it had a century ago, when it was 40,000 km in diameter. At the present rate of reduction, it could potentially become circular by 2040, although this is unlikely because of the distortion effect of the neighboring jet streams.[101] It is not known how long the spot will last, or whether the change is a result of normal fluctuations.[102]

According to a study by scientists at the University of California, Berkeley, between 1996 and 2006 the spot lost 15 percent of its diameter along its major axis. Xylar Asay-Davis, who was on the team that conducted the study, noted that the spot is not disappearing because "velocity is a more robust measurement because the clouds associated with the Red Spot are also strongly influenced by numerous other phenomena in the surrounding atmosphere."

Infrared data have long indicated that the Great Red Spot is colder (and thus, higher in altitude) than most of the other clouds on the planet;[103] the cloudtops of the GRS are about 8 km above the surrounding clouds. Furthermore, careful tracking of atmospheric features revealed the spot's counterclockwise circulation as far back as 1966 – observations dramatically confirmed by the first time-lapse movies from the *Voyager* flybys.[104] The spot is spatially confined by a modest eastward jet stream (prograde) to its south and a very strong westward (retrograde) one to its north.[105] Though winds around the edge of the spot peak at about 120 m/s (432 km/h), currents inside it seem stagnant, with little inflow or outflow.[106] The rotation period of the spot has decreased with

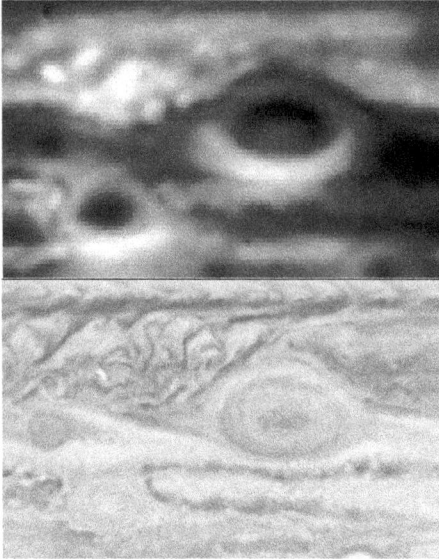

Figure 31: *An infrared image of GRS (top) and Oval BA (lower left) showing its cool center, taken by the ground based Very Large Telescope. An image made by the Hubble Space Telescope (bottom) is shown for comparison.*

time, perhaps as a direct result of its steady reduction in size. In 2010, astronomers imaged the GRS in the far infrared (from 8.5 to 24 µm) with a spatial resolution higher than ever before and found that its central, reddest region is warmer than its surroundings by between 3–4 K. The warm airmass is located in the upper troposphere in the pressure range of 200–500 mbar. This warm central spot slowly counter-rotates and may be caused by a weak subsidence of air in the center of GRS.[107]

The Great Red Spot's latitude has been stable for the duration of good observational records, typically varying by about a degree. Its longitude, however, is subject to constant variation.[108,109] Because Jupiter's visible features do not rotate uniformly at all latitudes, astronomers have defined three different systems for defining the longitude. System II is used for latitudes of more than 10°, and was originally based on the average rotation rate of the Great Red Spot of 9h 55m 42s.[110,111] Despite this, the spot has 'lapped' the planet in System II at least 10 times since the early 19th century. Its drift rate has changed dramatically over the years and has been linked to the brightness of the South Equatorial Belt, and the presence or absence of a South Tropical Disturbance.[112]

Figure 32: *Approximate size comparison of Earth superimposed on this Dec 29, 2000 image showing the Great Red Spot*

It is not known exactly what causes the Great Red Spot's reddish color. Theories supported by laboratory experiments suppose that the color may be caused by complex organic molecules, red phosphorus, or yet another sulfur compound. The GRS varies greatly in hue, from almost brick-red to pale salmon, or even white. The higher temperature of the reddest central region is the first evidence that the Spot's color is affected by environmental factors. The spot occasionally disappears from the visible spectrum, becoming evident only through the Red Spot Hollow, which is its niche in the South Equatorial Belt (SEB). The visibility of GRS is apparently coupled to the appearance of the SEB; when the belt is bright white, the spot tends to be dark, and when it is dark, the spot is usually light. The periods when the spot is dark or light occur at irregular intervals; in the 50 years from 1947 to 1997, the spot was darkest in the periods 1961–1966, 1968–1975, 1989–1990, and 1992–1993.[113] In November 2014, an analysis of data from NASA's Cassini mission revealed that the red color is likely a product of simple chemicals being broken apart by sunlight in the planet's upper atmosphere[114,115]

The Great Red Spot should not be confused with the Great Dark Spot, a feature observed near Jupiter's north pole in 2000 by the Cassini–Huygens spacecraft. A feature in the atmosphere of Neptune was also called the Great Dark Spot. The latter feature, imaged by *Voyager 2* in 1989, may have been an atmospheric

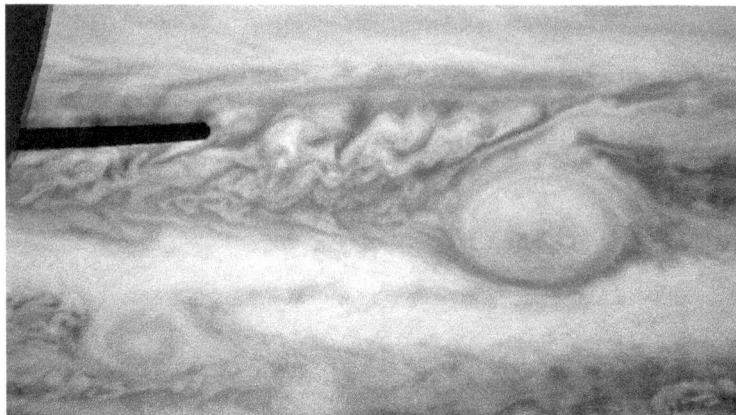

Figure 33: *Oval BA (left)*

hole rather than a storm. It was no longer present in 1994, although a similar spot had appeared farther to the north.[116]

Oval BA

Oval BA is a red storm in Jupiter's southern hemisphere similar in form to, though smaller than, the Great Red Spot (it is often affectionately referred to as "Red Spot Jr.", "Red Jr." or "The Little Red Spot"). A feature in the South Temperate Belt, Oval BA was first seen in 2000 after the collision of three small white storms, and has intensified since then.[117]

The formation of the three white oval storms that later merged into Oval BA can be traced to 1939, when the South Temperate Zone was torn by dark features that effectively split the zone into three long sections. Jovian observer Elmer J. Reese labeled the dark sections AB, CD, and EF. The rifts expanded, shrinking the remaining segments of the STZ into the white ovals FA, BC, and DE.[118] Ovals BC and DE merged in 1998, forming Oval BE. Then, in March 2000, BE and FA joined together, forming Oval BA. (see White ovals, below)

Oval BA slowly began to turn red in August 2005.[119] On February 24, 2006, Filipino amateur astronomer Christopher Go discovered the color change, noting that it had reached the same shade as the GRS. As a result, NASA writer Dr. Tony Phillips suggested it be called "Red Spot Jr." or "Red Jr."

In April 2006, a team of astronomers, believing that Oval BA might converge with the GRS that year, observed the storms through the Hubble Space Telescope. The storms pass each other about every two years, but the passings of 2002 and 2004 did not produce anything exciting. Dr. Amy Simon-Miller, of

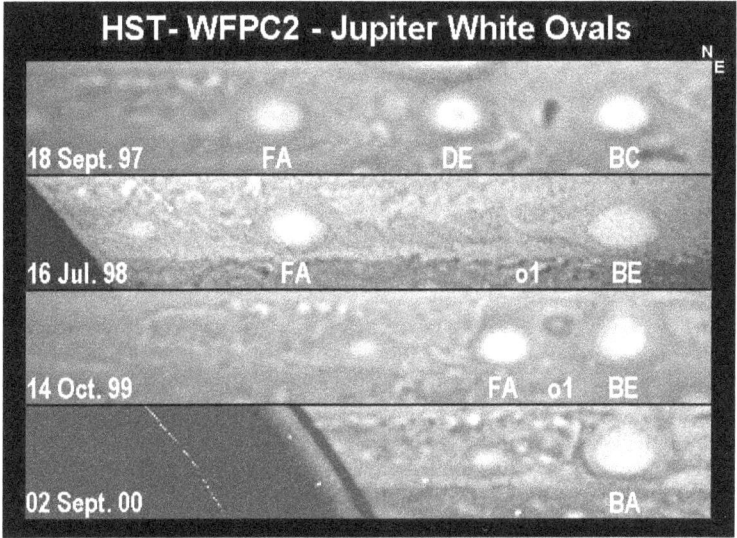

Figure 34: *Formation of Oval BA from three white ovals*

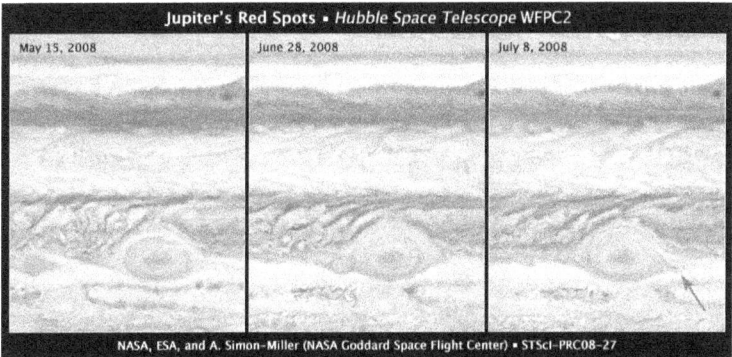

Figure 35: *Oval BA (bottom), Great Red Spot (top) and "Baby Red Spot" (middle) during a brief encounter in June, 2008*

the Goddard Space Flight Center, predicted the storms would have their closest passing on July 4, 2006. On July 20, the two storms were photographed
passing each other by the Gemini Observatory without converging.

Why Oval BA turned red is not understood. According to a 2008 study by Dr.
Santiago Pérez-Hoyos of the University of the Basque Country, the most likely
mechanism is "an upward and inward diffusion of either a colored compound
or a coating vapor that may interact later with high energy solar photons at
the upper levels of Oval BA." Some believe that small storms (and their corresponding white spots) on Jupiter turn red when the winds become powerful
enough to draw certain gases from deeper within the atmosphere which change
color when those gases are exposed to sunlight.

Oval BA is getting stronger according to observations made with the Hubble
Space Telescope in 2007. The wind speeds have reached 618 km/h; about
the same as in the Great Red Spot and far stronger than any of the progenitor
storms. As of July 2008, its size is about the diameter of Earth—approximately
half the size of the Great Red Spot.

Oval BA should not be confused with another major storm on Jupiter, the
South Tropical Little Red Spot (LRS) (nicknamed "the Baby Red Spot" by
NASA), which was destroyed by the GRS. The new storm, previously a white
spot in Hubble images, turned red in May 2008. The observations were led by
Imke de Pater of the University of California, at Berkeley, US. The Baby Red
Spot encountered the GRS in late June to early July 2008, and in the course of
a collision, the smaller red spot was shredded into pieces. The remnants of the
Baby Red Spot first orbited, then were later consumed by the GRS. The last
of the remnants with a reddish color to have been identified by astronomers
had disappeared by mid-July, and the remaining pieces again collided with
the GRS, then finally merged with the bigger storm. The remaining pieces
of the Baby Red Spot had completely disappeared by August 2008. During
this encounter Oval BA was present nearby, but played no apparent role in
destruction of the Baby Red Spot.

Storms and lightning

The storms on Jupiter are similar to thunderstorms on Earth. They reveal
themselves via bright clumpy clouds about 1000 km in size, which appear
from time to time in the belts' cyclonic regions, especially within the strong
westward (retrograde) jets.[120] In contrast to vortices, storms are short-lived
phenomena; the strongest of them may exist for several months, while the average lifetime is only 3–4 days. They are believed to be due mainly to moist
convection within Jupiter's troposphere. Storms are actually tall convective
columns (plumes), which bring the wet air from the depths to the upper part

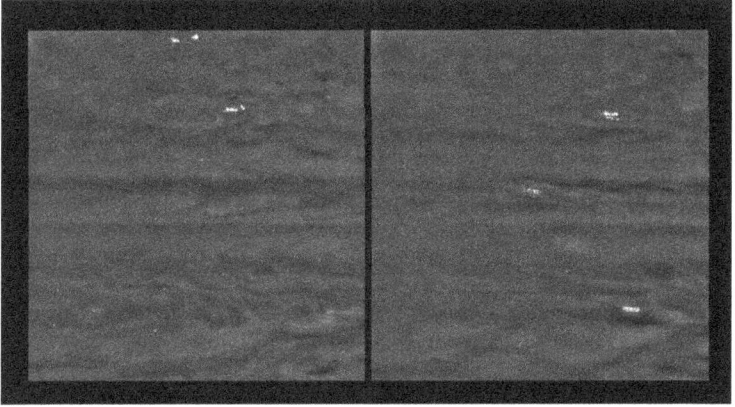

Figure 36: *Lightning on Jupiter's night side, imaged by the Galileo orbiter in 1997*

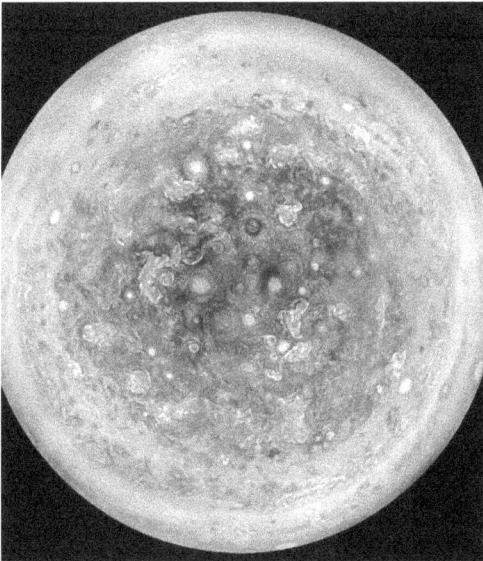

Figure 37: *Jupiter – southern storms – JunoCam*

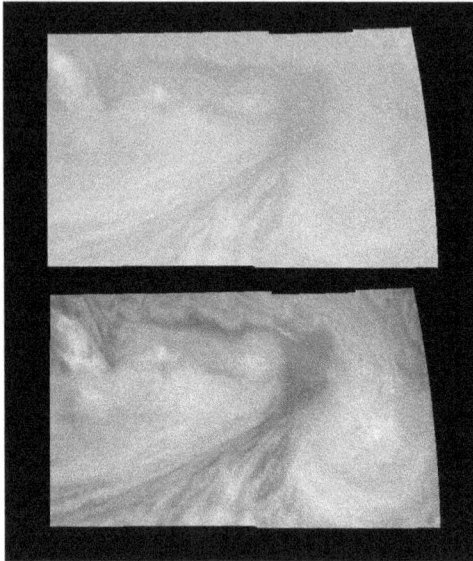

Figure 38: *False color image of an equatorial hot spot*

of the troposphere, where it condenses in clouds. A typical vertical extent of Jovian storms is about 100 km; as they extend from a pressure level of about 5–7 bar, where the base of a hypothetical water cloud layer is located, to as high as 0.2–0.5 bar.[121]

Storms on Jupiter are always associated with lightning. The imaging of the night–side hemisphere of Jupiter by *Galileo* and *Cassini* spacecraft revealed regular light flashes in Jovian belts and near the locations of the westward jets, particularly at 51°N, 56°S and 14°S latitudes.[122] On Jupiter lighting strikes are on average a few times more powerful than those on Earth. However, they are less frequent; the light power emitted from a given area is similar to that on Earth. A few flashes have been detected in polar regions, making Jupiter the second known planet after Earth to exhibit polar lightning.[123]

Every 15–17 years Jupiter is marked by especially powerful storms. They appear at 23°N latitude, where the strongest eastward jet, that can reach 150 m/s, is located. The last time such an event was observed was in March–June 2007. Two storms appeared in the northern temperate belt 55° apart in longitude. They significantly disturbed the belt. The dark material that was shed by the storms mixed with clouds and changed the belt's color. The storms moved with a speed as high as 170 m/s, slightly faster than the jet itself, hinting at the existence of strong winds deep in the atmosphere.

Disturbances

The normal pattern of bands and zones is sometimes disrupted for periods of time. One particular class of disruption are long-lived darkenings of the South Tropical Zone, normally referred to as "South Tropical Disturbances" (STD). The longest lived STD in recorded history was followed from 1901 until 1939, having been first seen by Percy B. Molesworth on February 28, 1901. It took the form of darkening over part of the normally bright South Tropical zone. Several similar disturbances in the South Tropical Zone have been recorded since then.[124]

Hot spots

One of the most mysterious features in the atmosphere of Jupiter are hot spots. In them the air is relatively free of clouds and heat can escape from the depths without much absorption. The spots look like bright spots in the infrared images obtained at the wavelength of about 5 μm. They are preferentially located in the belts, although there is a train of prominent hot spots at the northern edge of the Equatorial Zone. The Galileo Probe descended into one of those equatorial spots. Each equatorial spot is associated with a bright cloudy plume located to the west of it and reaching up to 10,000 km in size. Hot spots generally have round shapes, although they do not resemble vortexes.

The origin of hot spots is not clear. They can be either downdrafts, where the descending air is adiabatically heated and dried or, alternatively, they can be a manifestation of planetary scale waves. The latter hypotheses explains the periodical pattern of the equatorial spots.

Observational history

Early modern astronomers, using small telescopes, recorded the changing appearance of Jupiter's atmosphere. Their descriptive terms—belts and zones, brown spots and red spots, plumes, barges, festoons, and streamers—are still used.[125] Other terms such as vorticity, vertical motion, cloud heights have entered in use later, in the 20th century.

The first observations of the Jovian atmosphere at higher resolution than possible with Earth-based telescopes were taken by the *Pioneer 10* and *11* spacecraft. The first truly detailed images of Jupiter's atmosphere were provided by the *Voyagers*. The two spacecraft were able to image details at a resolution as low as 5 km in size in various spectra, and also able to create "approach movies" of the atmosphere in motion. The Galileo Probe, which suffered an antenna problem, saw less of Jupiter's atmosphere but at a better average resolution and a wider spectral bandwidth.

Figure 39: *Time-lapse sequence from the approach of Voyager 1 to Jupiter*

Today, astronomers have access to a continuous record of Jupiter's atmospheric activity thanks to telescopes such as Hubble Space Telescope. These show that the atmosphere is occasionally wracked by massive disturbances, but that, overall, it is remarkably stable. The vertical motion of Jupiter's atmosphere was largely determined by the identification of trace gases by ground-based telescopes. Spectroscopic studies after the collision of Comet Shoemaker–Levy 9 gave a glimpse of the Jupiter's composition beneath the cloud tops. The presence of diatomic sulfur (S_2) and carbon disulfide (CS_2) was recorded—the first detection of either in Jupiter, and only the second detection of S_2 in any astronomical object— together with other molecules such as ammonia (NH_3) and hydrogen sulfide (H_2S), while oxygen-bearing molecules such as sulfur dioxide were not detected, to the surprise of astronomers.[126]

The *Galileo* atmospheric probe, as it plunged into Jupiter, measured the wind, temperature, composition, clouds, and radiation levels down to 22 bar. However, below 1 bar elsewhere on Jupiter there is uncertainty in the quantities.

Figure 40: *A narrower view of Jupiter and the Great Red Spot as seen from Voyager 1 in 1979*

Great Red Spot studies

The first sighting of the GRS is often credited to Robert Hooke, who described a spot on the planet in May 1664; however, it is likely that Hooke's spot was in the wrong belt altogether (the North Equatorial Belt, versus the current location in the South Equatorial Belt). Much more convincing is Giovanni Cassini's description of a "permanent spot" in the following year.[127] With fluctuations in visibility, Cassini's spot was observed from 1665 to 1713.[128]

A minor mystery concerns a Jovian spot depicted around 1700 on a canvas by Donato Creti, which is exhibited in the Vatican.[129,130] It is a part of a series of panels in which different (magnified) heavenly bodies serve as backdrops for various Italian scenes, the creation of all of them overseen by the astronomer Eustachio Manfredi for accuracy. Creti's painting is the first known to depict the GRS as red. No Jovian feature was officially described as red before the late 19th century.

The present GRS was first seen only after 1830 and well-studied only after a prominent apparition in 1879. A 118-year gap separates the observations made after 1830 from its 17th-century discovery; whether the original spot dissipated and re-formed, whether it faded, or even if the observational record

Figure 41: *Hubble's Wide Field Camera 3*
took the GRS region at its smallest size ever.

was simply poor are unknown. The older spots had a short observational history and slower motion than that of the modern spot, which make their identity unlikely.

On February 25, 1979, when the *Voyager 1* spacecraft was 9.2 million kilometers from Jupiter it transmitted the first detailed image of the Great Red Spot back to Earth. Cloud details as small as 160 km across were visible. The colorful, wavy cloud pattern seen to the west (left) of the GRS is the spot's wake region, where extraordinarily complex and variable cloud motions are observed.[131]

White ovals

The white ovals that were to become Oval BA formed in 1939. They covered almost 90 degrees of longitude shortly after their formation, but contracted rapidly during their first decade; their length stabilized at 10 degrees or less after 1965.[132] Although they originated as segments of the STZ, they evolved to become completely embedded in the South Temperate Belt, suggesting that they moved north, "digging" a niche into the STB.[133] Indeed, much like the GRS, their circulations were confined by two opposing jet streams on their

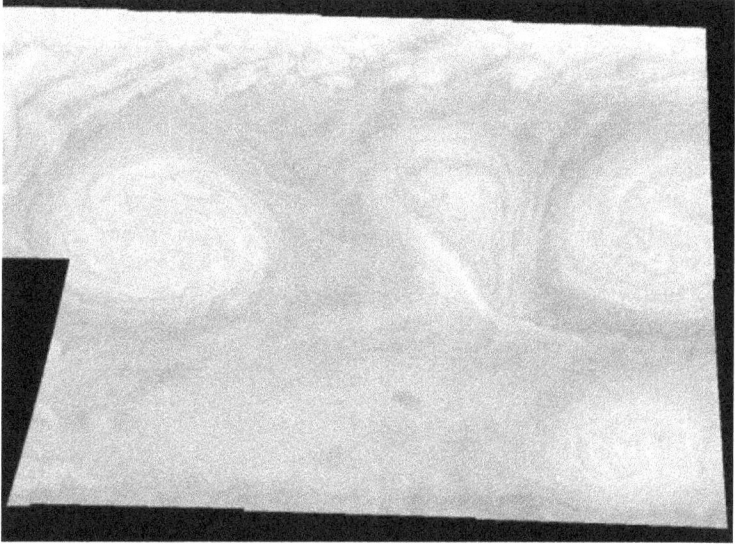

Figure 42: *The white ovals that later formed*
Oval BA, imaged by the Galileo orbiter in 1997

northern and southern boundaries, with an eastward jet to their north and a retrograde westward one to the south.

The longitudinal movement of the ovals seemed to be influenced by two factors: Jupiter's position in its orbit (they became faster at aphelion), and their proximity to the GRS (they accelerated when within 50 degrees of the Spot).[134] The overall trend of the white oval drift rate was deceleration, with a decrease by half between 1940 and 1990.[135]

During the *Voyager* fly-bys, the ovals extended roughly 9000 km from east to west, 5000 km from north to south, and rotated every five days (compared to six for the GRS at the time).[136]

Notes

<templatestyles src="Template:Refbegin/styles.css" />

a) ^ The scale height *sh* is defined as $sh = RT/(Mg_j)$, where $R = 8.31$ J/mol/K is the gas constant, $M \approx 0.0023$ kg/mol is the average molar mass in the Jovian atmosphere, T is temperature and $g_j \approx 25$ m/s^2 is the gravitational acceleration at the surface of Jupiter. As the temperature varies from 110 K in the tropopause up to 1000 K in the thermosphere, the scale height can assume values from 15 to 150 km.

b) ^ The *Galileo* atmospheric probe failed to measure the deep abundance of oxygen, because the water concentration continued to increase down to the pressure level of 22 bar, when it ceased operating. While the actually measured oxygen abundances are much lower than the solar value, the observed rapid increase of water content of the atmosphere with depth makes it highly likely that the deep abundance of oxygen indeed exceeds the solar value by a factor of about 3—much like other elements.[36]

c) ^ Various explanations of the overabundance of carbon, oxygen, nitrogen and other elements have been proposed. The leading one is that Jupiter captured a large number of icy planetesimals during the later stages of its accretion. The volatiles like noble gases are thought to have been trapped as clathrate hydrates in water ice.[36]

Cited sources

<templatestyles src="Template:Refbegin/styles.css" />

- Atreya, Sushil K.; Wong, M. H.; Owen, T. C.; Mahaffy, P. R.; Niemann, H. B.; de Pater, I.; Drossart, P.; Encrenaz, T. (October–November 1999). "A comparison of the atmospheres of Jupiter and Saturn: Deep atmospheric composition, cloud structure, vertical mixing, and origin". *Planetary and Space Science*. **47** (10–11): 1243–1262. Bibcode: 1999P&SS...47.1243A[137]. doi: 10.1016/S0032-0633(99)00047-1[138]. ISSN 0032-0633[139]. PMID 11543193[140].

- Atreya, Sushil K.; Mahaffy, P. R.; Niemann, H. B.; Wong, M. H.; Owen, T. C. (February 2003). "Composition and origin of the atmosphere of Jupiter—an update, and implications for the extrasolar giant planets". *Planetary and Space Science*. **51** (2): 105–112. Bibcode: 2003P&SS... 51..105A[141]. doi: 10.1016/S0032-0633(02)00144-7[142]. ISSN 0032-0633[139].

- Atreya, Sushil K.; Wong, Ah-San (2005). "Coupled Clouds and Chemistry of the Giant Planets — A Case for Multiprobes"[143] (PDF). *Space Science Reviews*. **116**: 121–136. Bibcode: 2005SSRv..116..121A[144]. doi: 10.1007/s11214-005-1951-5[145]. ISSN 0032-0633[139].

- Atreya, Sushil K.; Wong, Ah-San; Baines, K. H.; Wong, M. H.; Owen, T. C. (2005). "Jupiter's ammonia clouds—localized or ubiquitous?"[146] (PDF). *Planetary and Space Science*. **53** (5): 498–507. Bibcode: 2005P&SS...53..498A[147]. doi: 10.1016/j.pss.2004.04.002[148]. ISSN 0032-0633[139].

- Baines, Kevin H.; Simon-Miller, Amy A; Orton, Glenn S.; Weaver, Harold A.; Lunsford, Allen; Momary, Thomas W.; Spencer, John; Cheng, Andrew F.; Reuter, Dennis C. (12 October 2007). "Polar Lightning and Decadal-Scale Cloud Variability on Jupiter". *Science*. **318** (5848):

226–229. Bibcode: 2007Sci...318..226B[149]. doi: 10.1126/science. 1147912[150]. PMID 17932285[151].

• Beatty, J.K. (2002). "Jupiter's Shrinking Red Spot"[152]. *Sky and Telescope*. **103** (4): 24.

• Beebe, R. (1997). *Jupiter the Giant Planet* (2nd ed.). Washington: Smithsonian Books. ISBN 1-56098-685-9. OCLC 224014042[153].

• Bhardwaj, Anil; Gladstone, G. Randall (2000). "Auroral emissions of the giant planets"[154] (PDF). *Reviews of Geophysics*. **38** (3): 295–353. Bibcode: 2000RvGeo..38..295B[155]. doi: 10.1029/1998RG000046[156].

• Busse, F.H. (1976). "A simple model of convection in the Jovian atmosphere". *Icarus*. **29** (2): 255–260. Bibcode: 1976Icar...29..255B[157]. doi: 10.1016/0019-1035(76)90053-1[158].

• Encrenaz, Thérèse (February 2003). "ISO observations of the giant planets and Titan: what have we learnt?". *Planetary and Space Science*. **51** (2): 89–103. Bibcode: 2003P&SS...51...89E[159]. doi: 10.1016/S0032-0633(02)00145-9[160].

• Fletcher, Leigh N.; Orton,, G.S.; Mousis, O.; Yanamandra-Fisher, P.; et al. (2010). "Thermal structure and composition of Jupiter's Great Red Spot from high-resolution thermal imaging"[161] (PDF). *Icarus*. **208** (1): 306–328. Bibcode: 2010Icar..208..306F[162]. doi: 10.1016/j.icarus. 2010.01.005[163].

• Go, C.Y.; de Pater, I.; Wong, M.; Lockwood, S.; Marcus, P.; Asay-Davis, X.; Shetty, S. (2006). "Evolution Of The Oval Ba During 2004–2005". *Bulletin of the American Astronomical Society*. **38**: 495. Bibcode: 2006DPS....38.1102G[164].

• Graney, C. M. (2010). "Changes in the Cloud Belts of Jupiter, 1630–1664, as reported in the 1665 Astronomia Reformata of Giovanni Battista Riccioli". *Baltic Astronomy*. **19**: 266. arXiv: 1008.0566[165] ⊚. Bibcode: 2010BaltA..19..265G[166].

• Guillot, T. (1999). "A comparison of the interiors of Jupiter and Saturn". *Planetary and Space Science*. **47** (10–11): 1183–1200. arXiv: astro-ph/9907402[167] ⊚. Bibcode: 1999P&SS...47.1183G[168]. doi: 10.1016/S0032-0633(99)00043-4[169].

• Hammel, H.B.; Lockwood, G.W.; Mills, J.R.; Barnet, C.D. (1995). "Hubble Space Telescope Imaging of Neptune's Cloud Structure in 1994". *Science*. **268** (5218): 1740–1742. Bibcode: 1995Sci...268.1740H[170]. doi: 10.1126/science.268.5218.1740[171]. PMID 17834994[172].

• Heimpel, M.; Aurnou, J.; Wicht, J. (2005). "Simulation of equatorial and high-latitude jets on Jupiter in a deep convection model". *Nature*. **438** (7065): 193–196. Bibcode: 2005Natur.438..193H[173]. doi: 10.1038/nature04208[174]. PMID 16281029[175].

• Hockey, T. (1999). *Galileo's Planet: Observing Jupiter Before Photogra-phy*. Bristol, Philadelphia: Institute of Physics Publishing. ISBN 0-7503-0448-0. OCLC 39733730[176].

• Ingersoll, A.P.; Dowling, T.E.; Gierasch, P.J.; et al. (2004). "Dynam-ics of Jupiter's Atmosphere"[177] (PDF). In Bagenal, F.; Dowling, T.E.; McKinnon, W.B. *Jupiter: The Planet, Satellites and Magnetosphere*. Cambridge: Cambridge University Press. ISBN 0-521-81808-7.

• Ingersoll, A.P.; Cuzzi, J.N. (1969). "Dynamics of Jupiter's cloud bands". *Journal of the Atmospheric Sciences*. **26** (5): 981–985. Bibcode: 1969JAtS...26..981I[178]. doi: 10.1175/1520-0469(1969)026<0981:DOJCB>2.0.CO;2[179].

• Irwin, P. (2003). *Giant Planets of Our Solar System. Atmospheres, Com-position, and Structure*. Springer and Praxis. ISBN 978-3-540-00681-7.

• Kunde, V.G.; Flasar, F.M.; Jennings, D.E.; et al. (2004). "Jupiter's At-mospheric Composition from the Cassini Thermal Infrared Spectroscopy Experiment". *Science*. **305** (5690): 1582–1586. Bibcode: 2004Sci...305.1582K[180]. doi: 10.1126/science.1100240[181]. PMID 15319491[182].

• Low, F.J. (1966). "Observations of Venus, Jupiter, and Saturn at λ20 μ". *Astronomical Journal*. **71**: 391. Bibcode: 1966AJ.....71R.391L[183]. doi: 10.1086/110110[184].

• McKim, R.J. (1997). "P. B. Molesworth's discovery of the great South Tropical Disturbance on Jupiter, 1901". *Journal of the British Astronomi-cal Association*. **107** (5): 239–245. Bibcode: 1997JBAA..107..239M[185].

• Miller, Steve; Aylward, Alan; Millward, George (January 2005). "Giant Planet Ionospheres and Thermospheres: The Importance of Ion-Neutral Coupling". *Space Science Reviews*. **116** (1–2): 319–343. Bibcode: 2005SSRv..116..319M[186]. doi: 10.1007/s11214-005-1960-4[187].

• Noll, K.S.; McGrath, M.A.; Weaver, H.A.; Yelle, R.V.; et al. (1995). "HST Spectroscopic Observations of Jupiter Following the Impact of Comet Shoemaker-Levy 9". *Science*. **267** (5202): 1307–1313. Bibcode: 1995Sci...267.1307N[188]. doi: 10.1126/science.7871428[189]. PMID 7871428[190].

• Pearl, J. C.; Conrath, B. J.; Hanel, R. A.; Pirraglia, J. A.; Coustenis, A. (March 1990). "The albedo, effective temperature, and energy balance of Uranus, as determined from Voyager IRIS data". *Icarus*. **84** (1): 12–28. Bibcode: 1990Icar...84...12P[191]. doi: 10.1016/0019-1035(90)90155-3[192]. ISSN 0019-1035[193].

• Reese, E.J.; Solberg, H.G. (1966). "Recent measures of the latitude and longitude of Jupiter's red spot". *Icarus*. **5** (1–6): 266–273. Bibcode: 1966Icar....5..266R[194]. doi: 10.1016/0019-1035(66)90036-4[195].

• Ridpath, I. (1998). *Norton's Star Atlas and Reference Handbook* (19th ed.). Harlow: Addison Wesley Longman. p. 107. ISBN 0-582-35655-5.

• Rogers, J.H. (1995). *The Giant Planet Jupiter*. Cambridge: Cambridge University Press. ISBN 0-521-41008-8. OCLC 219591510[196].

• Rogers, J.H.; Metig, H.J. (2001). "Jupiter in 1998/99"[197] (PDF). *Journal of the British Astronomical Association*. **111** (6): 321–332. Bibcode: 2001JBAA..111..321R[198].

• Rogers, J.H. (2003). "Jupiter in 1999/2000. II: Infrared wavelengths"[199] (PDF). *Journal of the British Astronomical Association*. **113** (3): 136–140. Bibcode: 2003JBAA..113..136R[200].

• Rogers, J.H. (2008). "The accelerating circulation of Jupiter's Great Red Spot"[201] (PDF). *Journal of the British Astronomical Association*. **118** (1): 14–20. Bibcode: 2008JBAA..118...14R[202].

• Sanchez-Lavega, A.; Orton, G.S.; Morales R.; et al. (2001). "The Merger of Two Giant Anticyclones in the Atmosphere of Jupiter". *Icarus*. **149** (2): 491–495. Bibcode: 2001Icar..149..491S[203]. doi: 10.1006/icar. 2000.6548[204].

• Sanchez-Lavega, A.; Orton, G.S.; Hueso, S.; et al. (2008). "Depth of the strong Jovian jet from a planetary scale disturbance driven by storms". *Nature*. **451** (7177): 437–440. Bibcode: 2008Natur.451..437S[205]. doi: 10.1038/nature06533[206]. PMID 18216848[207].

• Seiff, A.; Kirk, D.B.; Knight, T.C.D.; et al. (1998). "Thermal structure of Jupiter's atmosphere near the edge of a 5-µm hot spot in the north equatorial belt". *Journal of Geophysical Research*. **103** (E10): 22857–22889. Bibcode: 1998JGR...10322857S[208]. doi: 10.1029/98JE01766[209].

• Smith, B.A.; Soderblom, L.A.; Johnson, T.V.; et al. (1979). "The Jupiter system through the eyes of Voyager 1". *Science*. **204** (4396): 951–957, 960–972. Bibcode: 1979Sci...204..951S[210]. doi: 10.1126/science.204.4396.951[211]. PMID 17800430[212].

• Stone, P.H. (1974). "On Jupiter's Rate of Rotation"[213] (PDF). *Journal of Atmospheric Sciences*. **31** (5): 1471–1472. Bibcode: 1974JAtS... 31.1471S[214]. doi: 10.1175/1520-0469(1974)031<1471:OJROR>2.0.CO; 2[215].

• Vasavada, A.R.; Showman, A. (2005). "Jovian atmospheric dynamics: An update after Galileo and Cassini". *Reports on Progress in Physics*. **68** (8): 1935–1996. Bibcode: 2005RPPh...68.1935V[216]. doi: 10.1088/0034-4885/68/8/R06[217].

• West, R.A.; Baines, K.H.; Friedson, A.J.; et al. (2004). "Jovian Clouds and Haze". In Bagenal, F.; Dowling, T.E.; McKinnon, W.B. *Jupiter: The Planet, Satellites and Magnetosphere*[218] (PDF). Cambridge: Cambridge University Press.

• Yelle, R.V.; Miller, S. (2004). "Jupiter's Thermosphere and Ionosphere"[219] (PDF). In Bagenal, F.; Dowling, T.E.; McKinnon, W.B. *Jupiter: The Planet, Satellites and Magnetosphere*. Cambridge: Cambridge University Press.

Further reading

- [Numerous authors] (1999). Beatty, Kelly J.; Peterson, Carolyn Collins; Chaiki, Andrew, eds. *The New Solar System* (4th ed.). Massachusetts: Sky Publishing Corporation. ISBN 0-933346-86-7. OCLC 39464951[220].
- Peek, Bertrand M. (1981). *The Planet Jupiter: The Observer's Handbook* (Revised ed.). London: Faber and Faber Limited. ISBN 0-571-18026-4. OCLC 8318939[221].
- Yang, Sarah (April 21, 2004). "Researcher predicts global climate change on Jupiter as giant planet's spots disappear"[222]. UC Berkeley News. Archived[223] from the original on 9 June 2007. Retrieved 2007-06-14.
- Youssef, Ashraf; Marcus, Philip S. (2003). "The dynamics of jovian white ovals from formation to merger". *Icarus.* **162** (1): 74–93. Bibcode: 2003Icar..162...74Y[224]. doi: 10.1016/S0019-1035(02)00060-X[225].
- Williams, Gareth P. (1975). "Jupiter's atmospheric circulation"[226] (PDF). *Nature.* **257** (5529): 778. Bibcode: 1975Natur.257..778W[227]. doi: 10.1038/257778a0[228].
- Williams, Gareth P. (1978). "Planetary Circulations: 1. Barotropic representation of Jovian and terrestrial turbulence"[229] (PDF). *Journal of the Atmospheric Sciences.* **35** (8): 1399–1426. Bibcode: 1978JAtS...35.1399W[230]. doi: 10.1175/1520-0469(1978)035<1399:PCBROJ>2.0.CO;2[231].
- Williams, Gareth P. (1985). "Jovian and comparative atmospheric modeling"[232] (PDF). *Advances in Geophysics.* Advances in Geophysics. **28A**: 381–429. Bibcode: 1985AdGeo..28..381W[233]. doi: 10.1016/S0065-2687(08)60231-9[234]. ISBN 978-0-12-018828-4.
- Williams, Gareth P. (1997). "Planetary vortices and Jupiter's vertical structure"[235] (PDF). *Journal of Geophysical Research.* **102** (E4): 9303–9308. Bibcode: 1997JGR...102.9303W[236]. doi: 10.1029/97JE00520[237].
- Williams, Gareth P. (1996). "Jovian Dynamics. Part I: Vortex stability, structure, and genesis"[238] (PDF). *Journal of the Atmospheric Sciences.* **53** (18): 2685–2734. Bibcode: 1996JAtS...53.2685W[239]. doi: 10.1175/1520-0469(1996)053<2685:JDPVSS>2.0.CO;2[240].
- Williams, Gareth P. (2002). "Jovian Dynamics.Part II: The genesis and equilibration of vortex sets"[241] (PDF). *Journal of the Atmospheric Sciences.* **59** (8): 1356–1370. Bibcode: 2002JAtS...59.1356W[242]. doi: 10.1175/1520-0469(2002)059<1356:JDPITG>2.0.CO;2[243].

- Williams, Gareth P. (2003). "Jovian Dynamics, Part III: Multiple, migrating, and equatorial jets"[244] (PDF). *Journal of the Atmospheric Sciences*. **60** (10): 1270–1296. Bibcode: 2003JAtS...60.1270W[245]. doi: 10.1175/1520-0469(2003)60<1270:JDPIMM>2.0.CO;2[246].
- Williams, Gareth P. (2003). "Super Circulations"[247] (PDF). *Bulletin of the American Meteorological Society*. **84** (9): 1190.
- Williams, Gareth P. (2003). "Barotropic instability and equatorial superrotation"[248] (PDF). *Journal of the Atmospheric Sciences*. **60** (17): 2136–2152. Bibcode: 2003JAtS...60.2136W[249]. doi: 10.1175/1520-0469(2003)060<2136:BIAES>2.0.CO;2[250].
- Williams, Gareth P. (2003). "Jet sets"[251] (PDF). *Journal of the Meteorological Society of Japan*. **81** (3): 439–476. doi: 10.2151/jmsj.81.439[252].
- Williams, Gareth P. (2006). "Equatorial Superrotation and Barotropic Instability: Static Stability Variants"[253] (PDF). *Journal of the Atmospheric Sciences*. **63** (5): 1548–1557. Bibcode: 2006JAtS...63.1548W[254]. doi: 10.1175/JAS3711.1[255].

External links

Wikimedia Commons has media related to <wbr />*Atmosphere of Jupiter* and <wbr />*Great Red Spot*.

- Planetary Society blog post[256] (2017-05-09) by Peter Rosén describing assembly of a video[257] of Jupiter's atmospheric activity from 19 December 2014 to 31 March 2015 from amateur astronomer images
- The Atmosphere[258]

<indicator name="featured-star"> ☆ </indicator>

Magnetosphere

Magnetosphere of Jupiter

Magnetosphere of Jupiter

Aurorae on the north pole of Jupiter
as viewed by Hubble

Discovery	
Discovered by	Pioneer 10
Discovery date	December 1973
Internal field[259]	
Radius of Jupiter	71,492 km
Magnetic moment	2.83×10^{20} T·m^3
Equatorial field strength	776.6 μT (7.766 G)
Dipole tilt	$\sim 10°$
Magnetic pole longitude	$\sim 159°$
Rotation period	9h 55m 29.7 ± 0.1s
Solar wind parameters[260]	
Speed	400 km/s

IMF strength	1 nT
Density	0.4 cm^{-3}
Magnetospheric parameters	
Type	Intrinsic
Bow shock distance	$\sim 82\ R_J$
Magnetopause distance	$50\text{--}100\ R_J$
Magnetotail length	up to $7000\ R_J$
Main ions	O^+, S^+ and H^+
Plasma sources	Io, solar wind, ionosphere
Mass loading rate	~ 1000 kg/s
Maximum plasma density	2000 cm^{-3}
Maximum particle energy	up to 100 MeV
Aurora	
Spectrum	radio, near-IR, UV and X-ray
Total power	100 TW
Radio emission frequencies	0.01–40 MHz

The **magnetosphere of Jupiter** is the cavity created in the solar wind by the planet's magnetic field. Extending up to seven million kilometers in the Sun's direction and almost to the orbit of Saturn in the opposite direction, Jupiter's magnetosphere is the largest and most powerful of any planetary magnetosphere in the Solar System, and by volume the largest known continuous structure in the Solar System after the heliosphere. Wider and flatter than the Earth's magnetosphere, Jupiter's is stronger by an order of magnitude, while its magnetic moment is roughly 18,000 times larger. The existence of Jupiter's magnetic field was first inferred from observations of radio emissions at the end of the 1950s and was directly observed by the *Pioneer 10* spacecraft in 1973.

Jupiter's internal magnetic field is generated by electrical currents in the planet's outer core, which is composed of liquid metallic hydrogen. Volcanic eruptions on Jupiter's moon Io eject large amounts of sulfur dioxide gas into space, forming a large torus around the planet. Jupiter's magnetic field forces the torus to rotate with the same angular velocity and direction as the planet. The torus in turn loads the magnetic field with plasma, in the process stretching it into a pancake-like structure called a magnetodisk. In effect, Jupiter's magnetosphere is shaped by Io's plasma and its own rotation, rather than by the solar wind like Earth's magnetosphere. Strong currents in the magnetosphere generate permanent auorae around the planet's poles and intense variable radio emissions, which means that Jupiter can be thought of as a very weak radio pulsar. Jupiter's aurorae have been observed in almost all parts of the electromagnetic spectrum, including infrared, visible, ultraviolet and soft X-rays.

The action of the magnetosphere traps and accelerates particles, producing intense belts of radiation similar to Earth's Van Allen belts, but thousands of times stronger. The interaction of energetic particles with the surfaces of Jupiter's largest moons markedly affects their chemical and physical properties. Those same particles also affect and are affected by the motions of the particles within Jupiter's tenuous planetary ring system. Radiation belts present a significant hazard for spacecraft and potentially to human space travellers.

Structure

Jupiter's magnetosphere is a complex structure comprising a bow shock, magnetosheath, magnetopause, magnetotail, magnetodisk, and other components. The magnetic field around Jupiter emanates from a number of different sources, including fluid circulation at the planet's core (the internal field), electrical currents in the plasma surrounding Jupiter and the currents flowing at the boundary of the planet's magnetosphere. The magnetosphere is embedded within the plasma of the solar wind, which carries the interplanetary magnetic field.[261]

Internal magnetic field

The bulk of Jupiter's magnetic field, like Earth's, is generated by an internal dynamo supported by the circulation of a conducting fluid in its outer core. But whereas Earth's core is made of molten iron and nickel, Jupiter's is composed of metallic hydrogen. As with Earth's, Jupiter's magnetic field is mostly a dipole, with north and south magnetic poles at the ends of a single magnetic axis. However, on Jupiter the north pole of the dipole is located in the planet's northern hemisphere and the south pole of the dipole lies in its southern hemisphere, opposite to the Earth, whose north pole lies in the southern hemisphere and south pole lies in the northern hemisphere.[262] Jupiter's field also has quadrupole, octupole and higher components, though they are less than one tenth as strong as the dipole component.[263]

The dipole is tilted roughly 10° from Jupiter's axis of rotation; the tilt is similar to that of the Earth (11.3°). Its equatorial field strength is about 776.6 μT (7.766 G), which corresponds to a dipole magnetic moment of about 2.83×10^{20} T·m^3. This makes Jupiter's magnetic field about 20 times stronger than Earth's, and its magnetic moment ~20,000 times larger.[264] Jupiter's magnetic field rotates at the same speed as the region below its atmosphere, with a period of 9 h 55 m. No changes in its strength or structure have been observed since the first measurements were taken by the Pioneer spacecraft in the mid-1970s.[265]

Size and shape

Jupiter's internal magnetic field prevents the solar wind, a stream of ionized particles emitted by the Sun, from interacting directly with its atmosphere, and instead diverts it away from the planet, effectively creating a cavity in the solar wind flow, called a magnetosphere, composed of a plasma different from that of the solar wind.[266] The Jovian (i.e. pertaining to Jupiter) magnetosphere is so large that the Sun and its visible corona would fit inside it with room to spare.[259] If one could see it from Earth, it would appear five times larger than the full moon in the sky despite being nearly 1700 times farther away.

As with Earth's magnetosphere, the boundary separating the denser and colder solar wind's plasma from the hotter and less dense one within Jupiter's magnetosphere is called the magnetopause. The distance from the magnetopause to the center of the planet is from 45 to 100 R_J (where R_J=71,492 km is the radius of Jupiter) at the subsolar point—the unfixed point on the surface at which the Sun would appear directly overhead to an observer. The position of the magnetopause depends on the pressure exerted by the solar wind, which in turn depends on solar activity.[267] In front of the magnetopause (at a distance from 80 to 130 R_J from the planet's center) lies the bow shock, a wake-like disturbance in the solar wind caused by its collision with the magnetosphere.[268,269] The region between the bow shock and magnetopause is called the magnetosheath.

At the opposite side of the planet, the solar wind stretches Jupiter's magnetic field lines into a long, trailing magnetotail, which sometimes extends well beyond the orbit of Saturn. The structure of Jupiter's magnetotail is similar to Earth's. It consists of two lobes (blue areas in the figure), with the magnetic field in the southern lobe pointing toward Jupiter, and that in the northern lobe pointing away from it. The lobes are separated by a thin layer of plasma called the tail current sheet (orange layer in the middle).[270] Like Earth's, the Jovian tail is a channel through which solar plasma enters the inner regions of the magnetosphere, where it is heated and forms the radiation belts at distances closer than 10 R_J from Jupiter.[271]

The shape of Jupiter's magnetosphere described above is sustained by the neutral sheet current (also known as the magnetotail current), which flows with Jupiter's rotation through the tail plasma sheet, the tail currents, which flow against Jupiter's rotation at the outer boundary of the magnetotail, and the magnetopause currents (or Chapman–Ferraro currents), which flow against rotation along the dayside magnetopause. These currents create the magnetic field that cancels the internal field outside the magnetosphere. They also interact substantially with the solar wind.[272]

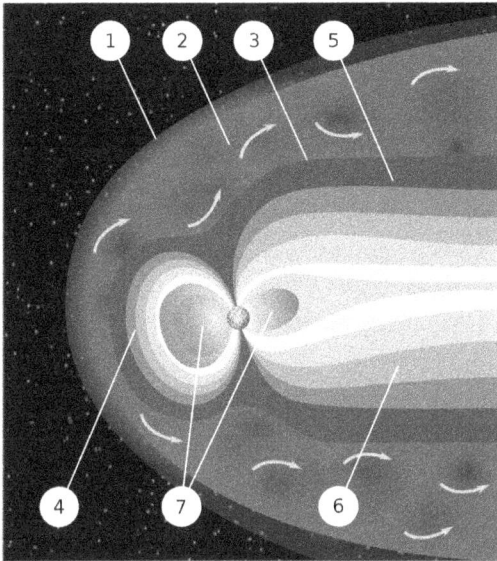

Figure 43: *An artist's concept of a magnetosphere, where plasmasphere (7) refers to the plasma torus and sheet*

Jupiter's magnetosphere is traditionally divided into three parts: the inner, middle and outer magnetosphere. The inner magnetosphere is located at distances closer than 10 R_J from the planet. The magnetic field within it remains approximately dipole, because contributions from the currents flowing in the magnetospheric equatorial plasma sheet are small. In the middle (between 10 and 40 R_J) and outer (further than 40 R_J) magnetospheres, the magnetic field is not a dipole, and is seriously disturbed by its interaction with the plasma sheet (see magnetodisk below).

Role of Io

Although overall the shape of Jupiter's magnetosphere resembles that of the Earth's, closer to the planet its structure is very different. Jupiter's volcanically active moon Io is a strong source of plasma in its own right, and loads Jupiter's magnetosphere with as much as 1,000 kg of new material every second.[273] Strong volcanic eruptions on Io emit huge amounts of sulfur dioxide, a major part of which is dissociated into atoms and ionized by the solar ultraviolet radiation, producing ions of sulfur and oxygen: S^+, O^+, S^{2+} and O^{2+}.[274]Wikipedia:Disputed statement These ions escape from the satellite's atmosphere and form the *Io plasma torus*: a thick and relatively cool ring of plasma encircling Jupiter, located near Io's orbit. The plasma temperature

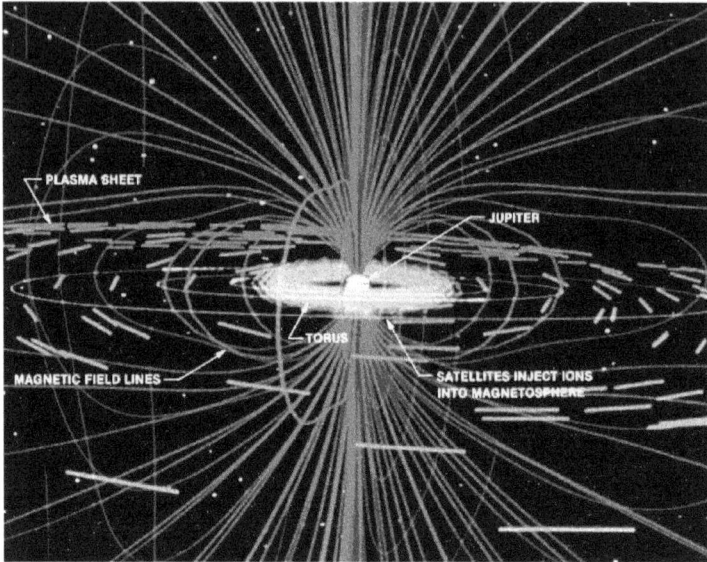

Figure 44: *Io's interaction with Jupiter's mag-*
netosphere. The Io plasma torus is in yellow.

within the torus is 10–100 eV (100,000–1,000,000 K), which is much lower than that of the particles in the radiation belts—10 keV (100 million K). The plasma in the torus is forced into co-rotation with Jupiter, meaning both share the same period of rotation. The Io torus fundamentally alters the dynamics of the Jovian magnetosphere.[275]

As a result of several processes—diffusion and interchange instability being the main escape mechanisms—the plasma slowly leaks away from Jupiter.[276] As the plasma moves further from the planet, the radial currents flowing within it gradually increase its velocity, maintaining co-rotation. These radial currents are also the source of the magnetic field's azimuthal component, which as a result bends back against the rotation. The particle number density of the plasma decreases from around 2,000 cm^{-3} in the Io torus to about 0.2 cm^{-3} at a distance of 35 R_J.[277] In the middle magnetosphere, at distances greater than 20 R_J from Jupiter, co-rotation gradually breaks down and the plasma begins to rotate more slowly than the planet. Eventually at the distances greater than 40 R_J (in the outer magnetosphere) this plasma escapes the magnetic field completely and leaves the magnetosphere through the magnetotail.[278] As cold, dense plasma moves outward, it is replaced by hot, low-density plasma (temperature 20 keV (200 million K) or higher) moving from the outer magnetosphere. This plasma, adiabatically heated as it approaches Jupiter,[279] forms the radiation belts in Jupiter's inner magnetosphere.

Magnetodisk

While Earth's magnetic field is roughly teardrop-shaped, Jupiter's is flatter, more closely resembling a disk, and "wobbles" periodically about its axis. The main reasons for this disk-like configuration are the centrifugal force from the co-rotating plasma and thermal pressure of hot plasma, both of which act to stretch Jupiter's magnetic field lines, forming a flattened pancake-like structure, known as the magnetodisk, at the distances greater than 20 R_J from the planet.[280] The magnetodisk has a thin current sheet at the middle plane, approximately near the magnetic equator. The magnetic field lines point away from Jupiter above the sheet and towards Jupiter below it. The load of plasma from Io greatly expands the size of the Jovian magnetosphere, because the magnetodisk creates an additional internal pressure which balances the pressure of the solar wind. In the absence of Io the distance from the planet to the magnetopause at the subsolar point would be no more than 42 R_J, whereas it is actually 75 R_J on average.

The configuration of the magnetodisk's field is maintained by the azimuthal ring current (not an analog of Earth's ring current), which flows with rotation through the equatorial plasma sheet.[281] The Lorentz force resulting from the interaction of this current with the planetary magnetic field creates a centripetal force, which keeps the co-rotating plasma from escaping the planet. The total ring current in the equatorial current sheet is estimated at 90–160 million amperes.[282]

Dynamics

Co-rotation and radial currents

The main driver of Jupiter's magnetosphere is the planet's rotation.[283] In this respect Jupiter is similar to a device called a Unipolar generator. When Jupiter rotates, its ionosphere moves relatively to the dipole magnetic field of the planet. Because the dipole magnetic moment points in the direction of the rotation, the Lorentz force, which appears as a result of this motion, drives negatively charged electrons to the poles, while positively charged ions are pushed towards the equator.[284] As a result, the poles become negatively charged and the regions closer to the equator become positively charged. Since the magnetosphere of Jupiter is filled with highly conductive plasma, the electrical circuit is closed through it. A current called the direct current[285] flows along the magnetic field lines from the ionosphere to the equatorial plasma sheet. This current then flows radially away from the planet within the equatorial plasma

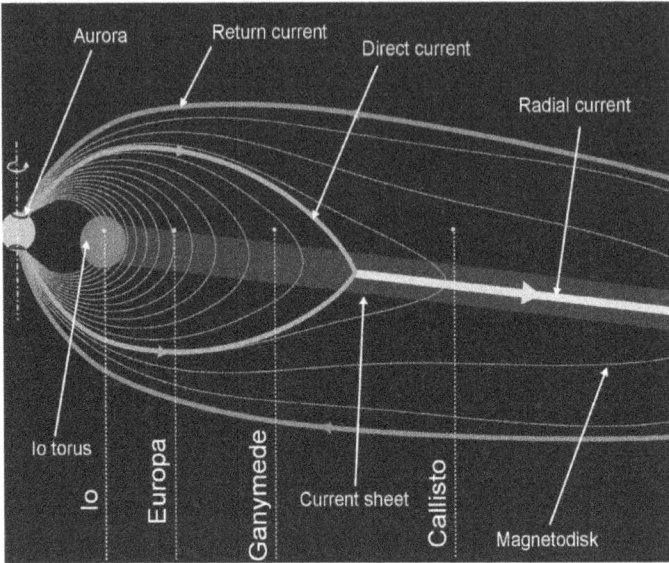

Figure 45: *The magnetic field of Jupiter and co-rotation enforcing currents*

sheet and finally returns to the planetary ionosphere from the outer reaches of the magnetosphere along the field lines connected to the poles. The currents that flow along the magnetic field lines are generally called field-aligned or Birkeland currents. The radial current interacts with the planetary magnetic field, and the resulting Lorentz force accelerates the magnetospheric plasma in the direction of planetary rotation. This is the main mechanism that maintains co-rotation of the plasma in Jupiter's magnetosphere.

The current flowing from the ionosphere to the plasma sheet is especially strong when the corresponding part of the plasma sheet rotates slower than the planet. As mentioned above, co-rotation breaks down in the region located between 20 and 40 R_J from Jupiter. This region corresponds to the magnetodisk, where the magnetic field is highly stretched.[286] The strong direct current flowing into the magnetodisk originates in a very limited latitudinal range of about 16 ± 1° from the Jovian magnetic poles. These narrow circular regions correspond to Jupiter's main auroral ovals. (See below.)[287] The return current flowing from the outer magnetosphere beyond 50 R_J enters the Jovian ionosphere near the poles, closing the electrical circuit. The total radial current in the Jovian magnetosphere is estimated at 60 million–140 million amperes.

The acceleration of the plasma into the co-rotation leads to the transfer of energy from the Jovian rotation to the kinetic energy of the plasma. In that

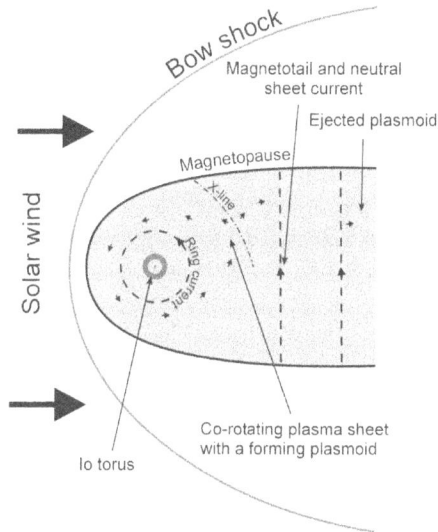

Figure 46: *The magnetosphere of Jupiter as viewed from above the north pole*

sense, the Jovian magnetosphere is powered by the planet's rotation, whereas the Earth's magnetosphere is powered mainly by the solar wind.

Interchange instability and reconnection

The main problem encountered in deciphering the dynamics of the Jovian magnetosphere is the transport of heavy cold plasma from the Io torus at 6 R_J to the outer magnetosphere at distances of more than 50 R_J. The precise mechanism of this process is not known, but it is hypothesized to occur as a result of plasma diffusion due to interchange instability. The process is similar to the Rayleigh-Taylor instability in hydrodynamics. In the case of the Jovian magnetosphere, centrifugal force plays the role of gravity; the heavy liquid is the cold and dense Ionian (i.e. pertaining to Io) plasma, and the light liquid is the hot, much less dense plasma from the outer magnetosphere. The instability leads to an exchange between the outer and inner parts of the magnetosphere of flux tubes filled with plasma. The buoyant empty flux tubes move towards the planet, while pushing the heavy tubes, filled with the Ionian plasma, away from Jupiter. This interchange of flux tubes is a form of magnetospheric turbulence.[288]

This highly hypothetical picture of the flux tube exchange was partly confirmed by the Galileo spacecraft, which detected regions of sharply reduced plasma density and increased field strength in the inner magnetosphere. These voids may correspond to the almost empty flux tubes arriving from the outer magnetosphere. In the middle magnetosphere, Galileo detected so-called injection events, which occur when hot plasma from the outer magnetosphere impacts the magnetodisk, leading to increased flux of energetic particles and a strengthened magnetic field.[289] No mechanism is yet known to explain the transport of cold plasma outward.

When flux tubes loaded with the cold Ionian plasma reach the outer magnetosphere, they go through a reconnection process, which separates the magnetic field from the plasma. The former returns to the inner magnetosphere in the form of flux tubes filled with hot and less dense plasma, while the latter are probably ejected down the magnetotail in the form of plasmoids—large blobs of plasma. The reconnection processes may correspond to the global reconfiguration events also observed by the Galileo spacecraft, which occurred regularly every 2–3 days. The reconfiguration events usually included rapid and chaotic variation of the magnetic field strength and direction, as well as abrupt changes in the motion of the plasma, which often stopped co-rotating and began flowing outward. They were mainly observed in the dawn sector of the night magnetosphere.[290] The plasma flowing down the tail along the open field lines is called the planetary wind.[291]

The reconnection events are analogues to the magnetic substorms in the Earth's magnetosphere. The difference seems to be their respective energy sources: terrestrial substorms involve storage of the solar wind's energy in the magnetotail followed by its release through a reconnection event in the tail's neutral current sheet. The latter also creates a plasmoid which moves down the tail.[292] Conversely, in Jupiter's magnetosphere the rotational energy is stored in the magnetodisk and released when a plasmoid separates from it.

Influence of the solar wind

Whereas the dynamics of Jovian magnetosphere mainly depend on internal sources of energy, the solar wind probably has a role as well,[293] particularly as a source of high-energy protons.[294] The structure of the outer magnetosphere shows some features of a solar wind-driven magnetosphere, including a significant dawn–dusk asymmetry. In particular, magnetic field lines in the dusk sector are bent in the opposite direction to those in the dawn sector. In addition, the dawn magnetosphere contains open field lines connecting to the magnetotail, whereas in the dusk magnetosphere, the field lines are closed. All these observations indicate that a solar wind driven reconnection process, known on Earth as the Dungey cycle, may also be taking place in the Jovian magnetosphere.

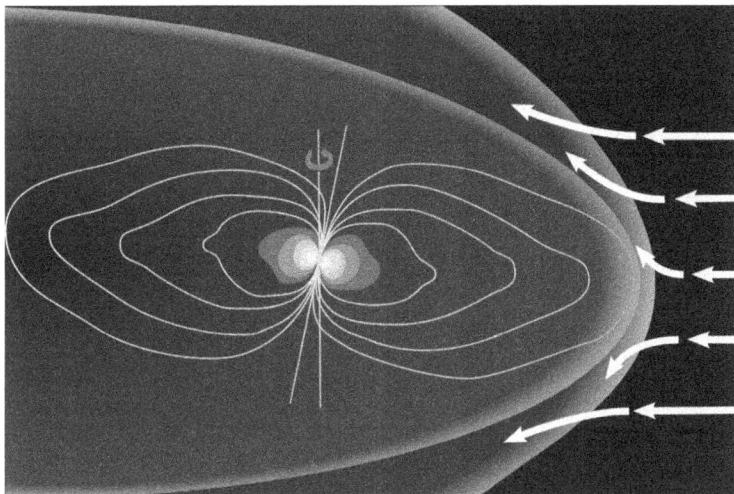

Figure 47: *Interactions between solar wind and Jovian magnetosphere*

The extent of the solar wind's influence on the dynamics of Jupiter's magnetosphere is currently unknown;[295] however, it could be especially strong at times of elevated solar activity.[296] The auroral radio, optical and X-ray emissions, as well as synchrotron emissions from the radiation belts all show correlations with solar wind pressure, indicating that the solar wind may drive plasma circulation or modulate internal processes in the magnetosphere.

Emissions

Aurorae

Jupiter demonstrates bright, persistent aurorae around both poles. Unlike Earth's aurorae, which are transient and only occur at times of heightened solar activity, Jupiter's aurorae are permanent, though their intensity varies from day to day. They consist of three main components: the main ovals, which are bright, narrow (less than 1000 km in width) circular features located at approximately $16°$ from the magnetic poles; the satellites' auroral spots, which correspond to the footprints of the magnetic field lines connecting Jupiter's ionosphere with those of its largest moons, and transient polar emissions situated within the main ovals.[297,298] Whereas the auroral emissions were detected in almost all parts of the electromagnetic spectrum from radio waves to X-rays (up to 3 keV), they are brightest in the mid-infrared (wavelength 3–4 μm and 7–14 μm) and deep ultraviolet spectral regions (wavelength 80–180 nm).[299]

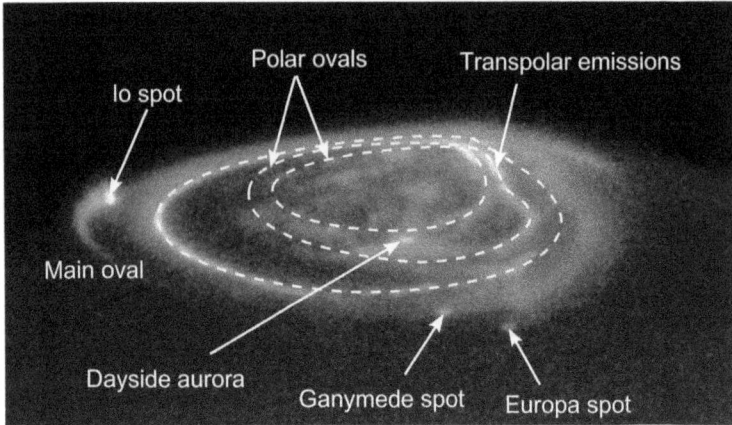

Figure 48: *Image of Jupiter's northern aurorae, showing the main auroral oval, the polar emissions, and the spots generated by the interaction with Jupiter's natural satellites*

The main ovals are the dominant part of the Jovian aurorae. They have stable shapes and locations, but their intensities are strongly modulated by the solar wind pressure—the stronger solar wind, the weaker the aurorae.[300] As mentioned above, the main ovals are maintained by the strong influx of electrons accelerated by the electric potential drops between the magnetodisk plasma and the Jovian ionosphere.[301] These electrons carry field aligned currents, which maintain the plasma's co-rotation in the magnetodisk. The potential drops develop because the sparse plasma outside the equatorial sheet can only carry a current of a limited strength without those currents. The precipitating electrons have energy in the range 10–100 keV and penetrate deep into the atmosphere of Jupiter, where they ionize and excite molecular hydrogen causing ultraviolet emission.[302] The total energy input into the ionosphere is 10–100 TW.[303] In addition, the currents flowing in the ionosphere heats it by the process known as Joule heating. This heating, which produces up to 300 TW of power, is responsible for the strong infrared radiation from the Jovian aurorae and partially for the heating of the thermosphere of Jupiter.[304]

Power emitted by Jovian aurorae in different parts of spectrum[305]

Emission	Jupiter	Io spot
Radio (KOM, <0.3 MHz)	~1 GW	?
Radio (HOM, 0.3–3 MHz)	~10 GW	?
Radio (DAM, 3–40 MHz)	~100 GW	0.1–1 GW (Io-DAM)
IR (hydrocarbons, 7–14 μm)	~40 TW	30–100 GW
IR (H_3^+, 3–4 μm)	4–8 TW	
Visible (0.385–1 μm)	10–100 GW	0.3 GW
UV (80–180 nm)	2–10 TW	~50 GW
X-ray (0.1–3 keV)	1–4 GW	?

Spots were found to correspond to three Galilean moons: Io, Europa and Ganymede.[306,307] They develop because the co-rotation of the plasma is slowed in the vicinity of moons. The brightest spot belongs to Io, which is the main source of the plasma in the magnetosphere (see above). The Ionian auroral spot is thought to be related to Alfvén currents flowing from the Jovian to Io-nian ionosphere. Europa's and Ganymede's spots are much dimmer, because these moons are weak plasma sources, because of sublimation of the water ice from their surfaces.[308]

Bright arcs and spots sporadically appear within the main ovals. These tran-sient phenomena are thought to be related to interaction with the solar wind. The magnetic field lines in this region are believed to be open or to map onto the magnetotail. The secondary ovals observed inside the main oval may be related to the boundary between open and closed magnetic field lines or to the polar cusps.[309] The polar auroral emissions are similar to those observed around Earth's poles: both appear when electrons are accelerated towards the planet by potential drops, during reconnection of solar magnetic field with that of the planet. The regions within both main ovals emit most of auroral X-rays. The spectrum of the auroral X-ray radiation consists of spectral lines of highly ionized oxygen and sulfur, which probably appear when energetic (hundreds of kiloelectronvolts) S and O ions precipitate into the polar atmosphere of Jupiter. The source of this precipitation remains unknown.[310]

Jupiter at radio wavelengths

Jupiter is a powerful source of radio waves in the spectral region stretching from several kilohertz to tens of megahertz. Radio waves with frequencies of less than about 0.3 MHz (and thus wavelengths longer than 1 km) are called the Jovian kilometric radiation or KOM. Those with frequencies in the interval

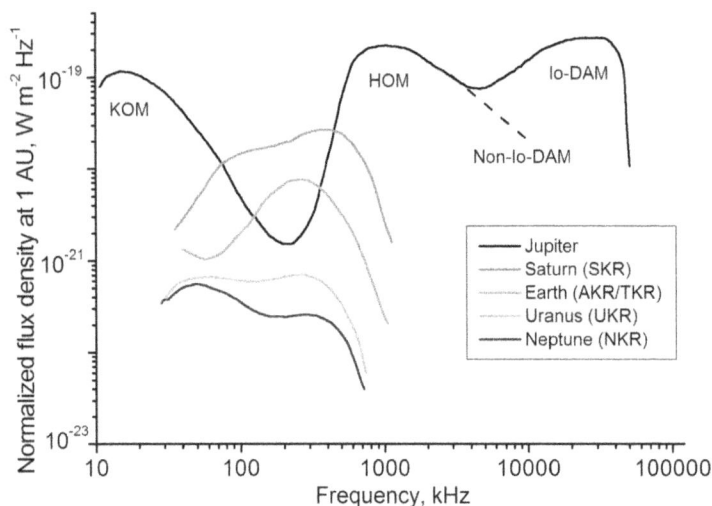

Figure 49: *The spectrum of Jovian radio emissions compared with spectra of four other magnetized planets, where (N,T,S,U)KR means (Neptunian, Terrestrial, Saturnian and Uranian) kilometric radiation*

of 0.3–3 MHz (with wavelengths of 100–1000 m) are called the hectometric radiation or HOM, while emissions in the range 3–40 MHz (with wavelengths of 10–100 m) are referred to as the decametric radiation or DAM. The latter radiation was the first to be observed from Earth, and its approximately 10-hour periodicity helped to identify it as originating from Jupiter. The strongest part of decametric emission, which is related to Io and to the Io–Jupiter current system, is called Io-DAM.[311,312]

The majority of these emissions are thought to be produced by a mechanism called Cyclotron Maser Instability, which develops close to the auroral regions, when electrons bounce back and forth between the poles. The electrons involved in the generation of radio waves are probably those carrying currents from the poles of the planet to the magnetodisk.[313] The intensity of Jovian radio emissions usually varies smoothly with time; however, Jupiter periodically emits short and powerful bursts (S bursts), which can outshine all other components. The total emitted power of the DAM component is about 100 GW, while the power of all other HOM/KOM components is about 10 GW. In comparison, the total power of Earth's radio emissions is about 0.1 GW.

Jupiter's radio and particle emissions are strongly modulated by its rotation, which makes the planet somewhat similar to a pulsar.[314] This periodical modulation is probably related to asymmetries in the Jovian magnetosphere, which are caused by the tilt of the magnetic moment with respect to the rotational axis as well as by high-latitude magnetic anomalies. The physics governing Jupiter's radio emissions is similar to that of radio pulsars. They differ only in the scale, and Jupiter can be considered a very small radio pulsar too. In addition, Jupiter's radio emissions strongly depend on solar wind pressure and, hence, on solar activity.

In addition to relatively long-wavelength radiation, Jupiter also emits synchrotron radiation (also known as the Jovian decimetric radiation or DIM radiation) with frequencies in the range of 0.1–15 GHz (wavelength from 3 m to 2 cm), which is the bremsstrahlung radiation of the relativistic electrons trapped in the inner radiation belts of the planet. The energy of the electrons that contribute to the DIM emissions is from 0.1 to 100 MeV,[315] while the leading contribution comes from the electrons with energy in the range 1–20 MeV.[316] This radiation is well understood and was used since the beginning of the 1960s to study the structure of the planet's magnetic field and radiation belts.[317] The particles in the radiation belts originate in the outer magnetosphere and are adiabatically accelerated, when they are transported to the inner magnetosphere.

Jupiter's magnetosphere ejects streams of high-energy electrons and ions (energy up to tens megaelectronvolts), which travel as far as Earth's orbit.[318] These streams are highly collimated and vary with the rotational period of the planet like the radio emissions. In this respect as well, Jupiter shows similarity to a pulsar.

Interaction with rings and moons

Jupiter's extensive magnetosphere envelops its ring system and the orbits of all four Galilean satellites. Orbiting near the magnetic equator, these bodies serve as sources and sinks of magnetospheric plasma, while energetic particles from the magnetosphere alter their surfaces. The particles sputter off material from the surfaces and create chemical changes via radiolysis.[319] The plasma's co-rotation with the planet means that the plasma preferably interacts with the moons' trailing hemispheres, causing noticeable hemispheric asymmetries.[319] In addition, the large internal magnetic fields of the moons contribute to the Jovian magnetic field.

Close to Jupiter, the planet's rings and small moons absorb high-energy particles (energy above 10 keV) from the radiation belts. This creates noticeable gaps in the belts' spatial distribution and affects the decimetric synchrotron radiation. In fact, the existence of Jupiter's rings was first hypothesized on the basis of data from the *Pioneer 11* spacecraft, which detected a sharp drop

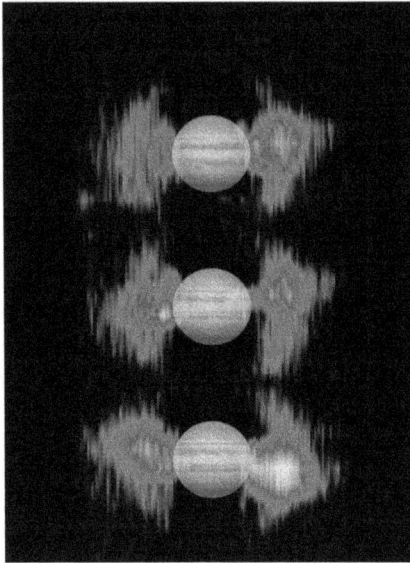

Figure 50: *Jupiter's variable radiation belts*

in the number of high-energy ions close to the planet.[320] The planetary magnetic field strongly influences the motion of sub-micrometer ring particles as well, which acquire an electrical charge under the influence of solar ultraviolet radiation. Their behavior is similar to that of co-rotating ions.[321] The resonant interaction between the co-rotation and the orbital motion is thought to be responsible for the creation of Jupiter's innermost halo ring (located between 1.4 and 1.71 R_J), which consists of sub-micrometer particles on highly inclined and eccentric orbits.[322] The particles originate in the main ring; however, when they drift toward Jupiter, their orbits are modified by the strong 3:2 Lorentz resonance located at 1.71 R_J, which increases their inclinations and eccentricities.[323] Another 2:1 Lorentz resonance at 1.4 Rj defines the inner boundary of the halo ring.[324]

All Galilean moons have thin atmospheres with surface pressures in the range 0.01–1 nbar, which in turn support substantial ionospheres with electron densities in the range of 1,000–10,000 cm^{-3}.[325] The co-rotational flow of cold magnetospheric plasma is partially diverted around them by the currents induced in their ionospheres, creating wedge-shaped structures known as Alfvén wings.[326] The interaction of the large moons with the co-rotational flow is similar to the interaction of the solar wind with the non-magnetized planets like Venus, although the co-rotational speed is usually subsonic (the speeds vary from 74 to 328 km/s), which prevents the formation of a bow shock.[327] The

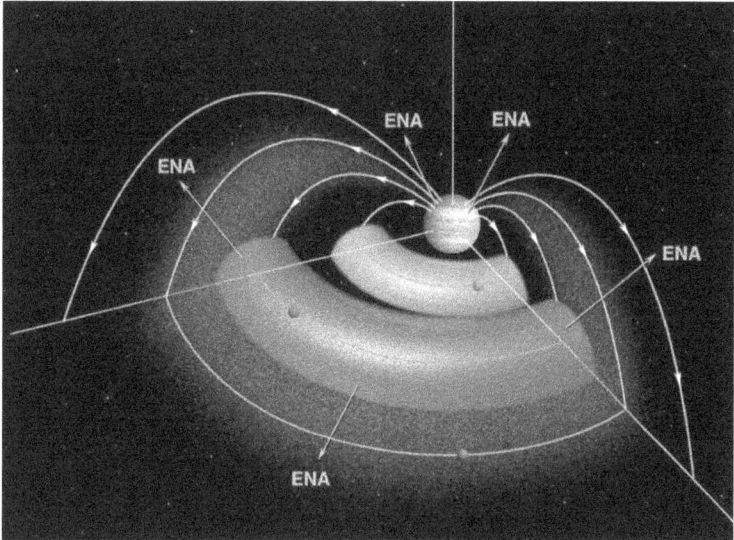

Figure 51: *Plasma tori created by Io and Europa*

pressure from the co-rotating plasma continuously strips gases from the moons' atmospheres (especially from that of Io), and some of these atoms are ionized and brought into co-rotation. This process creates gas and plasma tori in the vicinity of moons' orbits with the Ionian torus being the most prominent. In effect, the Galilean moons (mainly Io) serve as the principal plasma sources in Jupiter's inner and middle magnetosphere. Meanwhile, the energetic particles are largely unaffected by the Alfvén wings and have free access to the moons' surfaces (except Ganymede's).[328]

The icy Galilean moons, Europa, Ganymede and Callisto, all generate induced magnetic moments in response to changes in Jupiter's magnetic field. These varying magnetic moments create dipole magnetic fields around them, which act to compensate for changes in the ambient field. The induction is thought to take place in subsurface layers of salty water, which are likely to exist in all of Jupiter's large icy moons. These underground oceans can potentially harbor life, and evidence for their presence was one of the most important discoveries made in the 1990s by spacecraft.[329]

The interaction of the Jovian magnetosphere with Ganymede, which has an intrinsic magnetic moment, differs from its interaction with the non-magnetized moons. Ganymede's internal magnetic field carves a cavity inside Jupiter's magnetosphere with a diameter of approximately two Ganymede diameters, creating a mini-magnetosphere within Jupiter's magnetosphere. Ganymede's magnetic field diverts the co-rotating plasma flow around its magnetosphere. It also protects the moon's equatorial regions, where the field lines are closed, from energetic particles. The latter can still freely strike Ganymede's poles, where the field lines are open.[330] Some of the energetic particles are trapped near the equator of Ganymede, creating mini-radiation belts.[331] Energetic electrons entering its thin atmosphere are responsible for the observed Ganymedian polar aurorae.

Charged particles have a considerable influence on the surface properties of Galilean moons. Plasma originating from Io carries sulfur and sodium ions farther from the planet,[332] where they are implanted preferentially on the trailing hemispheres of Europa and Ganymede.[319] On Callisto however, for unknown reasons, sulfur is concentrated on the leading hemisphere.[333] Plasma may also be responsible for darkening the moons' trailing hemispheres (again, except Callisto's). Energetic electrons and ions, with the flux of the latter being more isotropic, bombard surface ice, sputtering atoms and molecules off and causing radiolysis of water and other chemical compounds. The energetic particles break water into oxygen and hydrogen, maintaining the thin oxygen atmospheres of the icy moons (since the hydrogen escapes more rapidly). The compounds produced radiolytically on the surfaces of Galilean moons also include ozone and hydrogen peroxide. If organics or carbonates are present, carbon dioxide, methanol and carbonic acid can be produced as well. In the presence of sulfur, likely products include sulfur dioxide, hydrogen disulfide and sulfuric acid.[319] Oxidants produced by radiolysis, like oxygen and ozone, may be trapped inside the ice and carried downward to the oceans over geologic time intervals, thus serving as a possible energy source for life.

Discovery

The first evidence for the existence of Jupiter's magnetic field came in 1955, with the discovery of the decametric radio emission or DAM.[334] As the DAM's spectrum extended up to 40 MHz, astronomers concluded that Jupiter must possess a magnetic field with a strength of about 1 milliteslas (10 gauss).[335]

In 1959, observations in the microwave part of the electromagnetic (EM) spectrum (0.1–10 GHz) led to the discovery of the Jovian decimetric radiation (DIM) and the realization that it was synchrotron radiation emitted by relativistic electrons trapped in the planet's radiation belts.[336] These synchrotron emissions were used to estimate the number and energy of the electrons around Jupiter and led to improved estimates of the magnetic moment and its tilt.

By 1973 the magnetic moment was known within a factor of two, whereas the tilt was correctly estimated at about 10°. The modulation of Jupiter's DAM by Io (the so-called Io-DAM) was discovered in 1964, and allowed Jupiter's

Figure 52: *Pioneer 10 provided the first in situ and definitive discovery of the Jovian magnetosphere*

rotation period to be precisely determined.[337] The definitive discovery of the Jovian magnetic field occurred in December 1973, when the Pioneer 10 spacecraft flew near the planet.[338,339]

Exploration after 1970

As of 2009 a total of eight spacecraft have flown around Jupiter and all have contributed to the present knowledge of the Jovian magnetosphere. The first space probe to reach Jupiter was Pioneer 10 in December 1973, which passed within 2.9 R_J from the center of the planet. Its twin Pioneer 11 visited Jupiter a year later, traveling along a highly inclined trajectory and approaching the planet as close as 1.6 R_J.

Pioneer provided the best coverage available of the inner magnetic field. The level of radiation at Jupiter was ten times more powerful than *Pioneer*'s designers had predicted, leading to fears that the probe would not survive; however, with a few minor glitches, it managed to pass through the radiation belts, saved in large part by the fact that Jupiter's magnetosphere had "wobbled" slightly upward at that point, moving away from the spacecraft. However, Pioneer 11 did lose most images of Io, as the radiation had caused its imaging photo

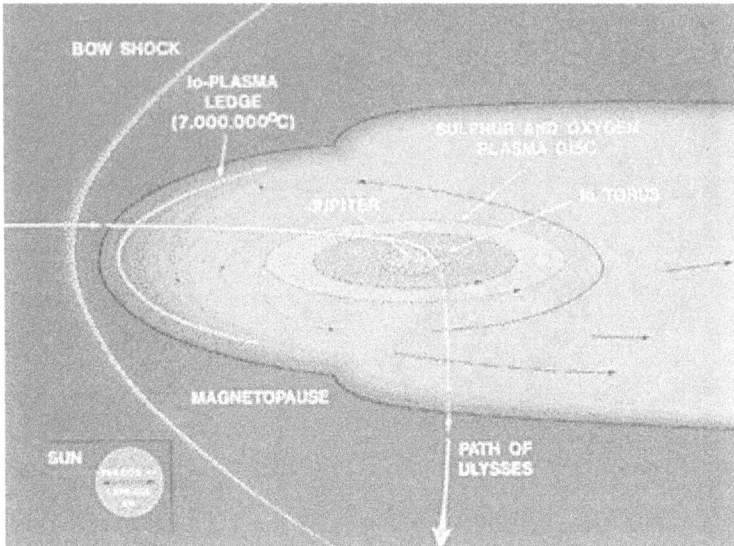

Figure 53: *The path of the Ulysses spacecraft through the magnetosphere of Jupiter in 1992*

Figure 54: *Galileo orbiter's magnetometer instrument*

polarimeter to receive a number of spurious commands. The subsequent and far more technologically advanced *Voyager* spacecraft had to be redesigned to cope with the massive radiation levels.[340]

Voyagers 1 and 2 arrived at Jupiter in 1979–1980 and traveled almost in its equatorial plane. Voyager 1, which passed within 5 R_J from the planet's center, was first to encounter the Io plasma torus. Voyager 2 passed within 10 R_J and discovered the current sheet in the equatorial plane. The next probe to approach Jupiter was Ulysses in 1992, which investigated the planet's polar magnetosphere.

The Galileo spacecraft, which orbited Jupiter from 1995 to 2003, provided a comprehensive coverage of Jupiter's magnetic field near the equatorial plane at distances up to 100 R_J. The regions studied included the magnetotail and the dawn and dusk sectors of the magnetosphere. While Galileo successfully survived in the harsh radiation environment of Jupiter, it still experienced a few technical problems. In particular, the spacecraft's gyroscopes often exhibited increased errors. Several times electrical arcs occurred between rotating and non-rotating parts of the spacecraft, causing it to enter safe mode, which led to total loss of the data from the 16th, 18th and 33rd orbits. The radiation also caused phase shifts in Galileo's ultra-stable quartz oscillator.[341]

When the Cassini spacecraft flew by Jupiter in 2000, it conducted coordinated measurements with Galileo. New Horizons passed close to Jupiter in 2007, carrying out a unique investigation of the Jovian magnetotail, traveling as far as 2500 R_J along its length.[342] In July 2016 Juno was inserted into Jupiter orbit, its scientific objectives include exploration of Jupiter's polar magnetosphere. The coverage of Jupiter's magnetosphere remains much poorer than for Earth's magnetic field. Further study is important to further understand the Jovian magnetosphere's dynamics.

In 2003, NASA conducted a conceptual study called "Human Outer Planets Exploration" (HOPE) regarding the future human exploration of the outer solar system. The possibility was mooted of building a surface base on Callisto, because of the low radiation levels at the moon's distance from Jupiter and its geological stability. Callisto is the only one of Jupiter's Galilean satellites for which human exploration is feasible. The levels of ionizing radiation on Io, Europa and Ganymede are inimical to human life, and adequate protective measures have yet to be devised.[343]

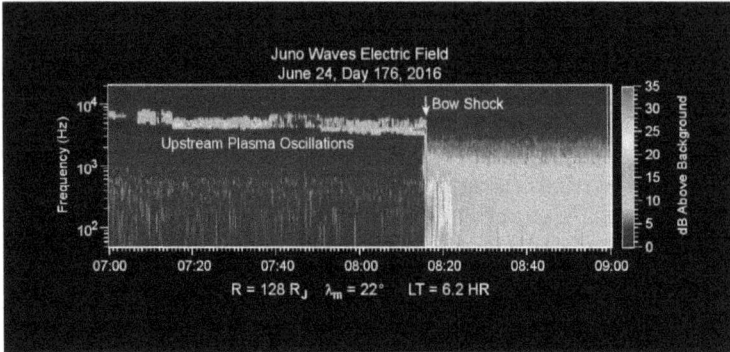

Figure 55: *Waves data as Juno crosses the Jovian bow shock (June 2016)*

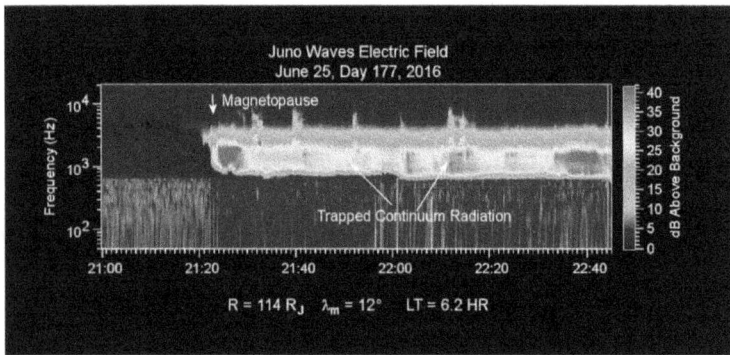

Figure 56: *Waves data as Juno enters magnetopause (June 2016)*

Exploration after 2010

The *Juno* New Frontiers mission to Jupiter was launched in 2011 and arrived at Jupiter in 2016. It includes a suite of instruments designed to better understand the magnetosphere, including a Magnetometer on Juno instrument as well as other devices such as a detector for Plasma and Radio fields called *Waves*.

The Jovian Auroral Distributions Experiment (JIRAM) instrument should also help to understand the magnetosphere.

<templatestyles src="Template:Quote/styles.css"/>

A primary objective of the Juno mission is to explore the polar magneto-sphere of Jupiter. While Ulysses briefly attained latitudes of ~48 degrees, this was at relatively large distances from Jupiter (~8.6 RJ). Hence, the polar magnetosphere of Jupiter is largely uncharted territory and, in particular, the auroral acceleration region has never been visited. ...

—*A Wave Investigation for the Juno Mission to Jupiter*

Juno revealed a planetary magnetic field rich in spatial variation, possibly due to a relatively large dynamo radius. The most surprising observation until late 2017 was the absence of the expected magnetic signature of intense field aligned currents (Birkeland currents) associated with the main aurora.[344]

Cited sources

<templatestyles src="Template:Refbegin/styles.css" />

- Bhardwaj, A.; Gladstone, G.R. (2000). "Auroral emissions of the giant planets"[345] (PDF). *Reviews of Geophysics*. **38** (3): 295–353. Bibcode: 2000RvGeo..38..295B[346]. doi: 10.1029/1998RG000046[347].
- Blanc, M.; Kallenbach, R.; Erkaev, N. V. (2005). "Solar System magnetospheres". *Space Science Reviews*. **116** (1–2): 227–298. Bibcode: 2005SSRv..116..227B[348]. doi: 10.1007/s11214-005-1958-y[349].
- Bolton, S.J.; Janssen, M.; et al. (2002). "Ultra-relativistic electrons in Jupiter's radiation belts"[350]. *Nature*. **415** (6875): 987–991. Bibcode: 2002Natur.415..987B[351]. doi: 10.1038/415987a[352]. PMID 11875557[353].
- Burke, B.F.; Franklin, K. L. (1955). "Observations of a variable radio source associated with the planet Jupiter". *Journal of Geophysical Research*. **60** (2): 213–217. Bibcode: 1955JGR....60..213B[354]. doi: 10.1029/JZ060i002p00213[355].
- Burns, J.A.; Simonelli, D. P.; Showalter; Hamilton; Porco; Throop; Esposito (2004). "Jupiter's ring-moon system"[356] (PDF). In Bagenal, F.; et al. *Jupiter: The Planet, Satellites and Magnetosphere*. Cambridge University Press. p. 241. Bibcode: 2004jpsm.book..241B[357]. ISBN 0-521-81808-7.
- Clarke, J.T.; Ajello, J.; et al. (2002). "Ultraviolet emissions from the magnetic footprints of Io, Ganymede and Europa on Jupiter"[358] (PDF). *Nature*. **415** (6875): 997–1000. doi: 10.1038/415997a[359]. PMID 11875560[360].
- Cooper, J. F.; Johnson, R. E.; et al. (2001). "Energetic ion and electron irradiation of the icy Galilean satellites"[361] (PDF). *Icarus*. **139** (1): 133–159. Bibcode: 2001Icar..149..133C[362]. doi: 10.1006/icar. 2000.6498[363]. Archived from the original[364] (PDF) on 2009-02-25.
- Cowley, S.W. H.; Bunce, E. J. (2001). "Origin of the main auroral oval in Jupiter's coupled magnetosphere–ionosphere system". *Planetary and Space Science*. **49** (10–11): 1067–66. Bibcode: 2001P&SS... 49.1067C[365]. doi: 10.1016/S0032-0633(00)00167-7[366].

• Cowley, S.W. H.; Bunce, E. J. (2003). "Modulation of Jovian middle
  magnetosphere currents and auroral precipitation by solar wind-induced
  compressions and expansions of the magnetosphere: initial response and
  steady state". *Planetary and Space Science.* **51** (1): 31–56. Bibcode:
  2003P&SS...51...31C[367]. doi: 10.1016/S0032-0633(02)00130-7[368].

• Drake, F. D.; Hvatum, S. (1959). "Non-thermal microwave radiation
  from Jupiter". *Astronomical Journal.* **64**: 329. Bibcode: 1959AJ.....64S.
  329D[369]. doi: 10.1086/108047[370].

• Elsner, R. F.; Ramsey, B. D.; et al. (2005). "X-ray probes of magnetospheric interactions with Jupiter's auroral zones, the Galilean satellites,
  and the Io plasma torus"[371] (PDF). *Icarus.* **178** (2): 417–428. Bibcode:
  2005Icar..178..417E[372]. doi: 10.1016/j.icarus.2005.06.006[373].

• Fieseler, P.D.; Ardalan, S. M.; et al. (2002). "The radiation effects on
  Galileo spacecraft systems at Jupiter"[374] (PDF). *Nuclear Science.* **49**
  (6): 2739–58. Bibcode: 2002ITNS...49.2739F[375]. doi: 10.1109/TNS.
  2002.805386[376]. Archived from the original[377] (PDF) on 2011-07-19.

• Hill, T. W.; Dessler, A. J. (1995). "Space physics and astronomy
  converge in exploration of Jupiter's Magnetosphere"[378]. *Earth
  in Space.* **8** (32): 6. Bibcode: 1995EOSTr..76..313H[379]. doi:
  10.1029/95EO00190[380]. Archived from the original[381] on 1997-05-01.

• Hibbitts, C.A.; McCord, T.B.; Hansen, T.B. (2000). "Distribution of
  CO_2 and SO_2 on the surface of Callisto". *Journal of Geophysical Research.* **105** (E9): 22,541–557. Bibcode: 2000JGR...10522541H[382]. doi:
  10.1029/1999JE001101[383].

• Johnson, R.E.; Carlson, R.V.; et al. (2004). "Radiation Effects on the
  Surfaces of the Galilean Satellites"[384] (PDF). In Bagenal, F.; et al. *Jupiter:
  The Planet, Satellites and Magnetosphere.* Cambridge University Press.
  ISBN 0-521-81808-7.

• Khurana, K.K.; Kivelson, M. G.; et al. (2004). "The configuration
  of Jupiter's magnetosphere"[385] (PDF). In Bagenal, F.; Dowling, T.E.;
  McKinnon, W.B. *Jupiter: The Planet, Satellites and Magnetosphere.*
  Cambridge University Press. ISBN 0-521-81808-7.

• Kivelson, M.G. (2005). "The current systems of the Jovian magnetosphere and ionosphere and predictions for Saturn"[386] (PDF). *Space
  Science Reviews.* Springer. **116** (1–2): 299–318. Bibcode: 2005SSRv..
  116..299K[387]. doi: 10.1007/s11214-005-1959-x[388].

• Kivelson, M.G.; Bagenal, F.; et al. (2004). "Magnetospheric interactions
  with satellites"[389] (PDF). In Bagenal, F.; Dowling, T.E.; McKinnon, W.B.
  Jupiter: The Planet, Satellites and Magnetosphere. Cambridge University
  Press. ISBN 0-521-81808-7.

• Krupp, N.; Vasyliunas, V.M.; et al. (2004). "Dynamics of the Jovian
  Magnetosphere"[390] (PDF). In Bagenal, F.; et al. *Jupiter: The Planet,*

Satellites and Magnetosphere. Cambridge University Press. ISBN 0-521-81808-7.

- Krupp, N. (2007). "New surprises in the largest magnetosphere of Our Solar System". *Science*. **318** (5848): 216–217. Bibcode: 2007Sci... 318..216K[391]. doi: 10.1126/science.1150448[392]. PMID 17932281[393].
- Miller, Steve; Aylward, Alan; Millward, George (January 2005). "Giant Planet Ionospheres and Thermospheres: The Importance of Ion-Neutral Coupling". *Space Science Reviews*. **116** (1–2): 319–343. Bibcode: 2005SSRv..116..319M[394]. doi: 10.1007/s11214-005-1960-4[395].
- Nichols, J. D.; Cowley, S. W. H.; McComas, D. J. (2006). "Magnetopause reconnection rate estimates for Jupiter's magnetosphere based on interplanetary measurements at ∼5 AU"[396]. *Annales Geophysicae*. **24** (1): 393–406. Bibcode: 2006AnGeo..24..393N[397]. doi: 10.5194/angeo-24-393-2006[398].
- Palier, L.; Prangé, Renée (2001). "More about the structure of the high latitude Jovian aurorae". *Planetary and Space Science*. **49** (10–11): 1159–73. Bibcode: 2001P&SS...49.1159P[399]. doi: 10.1016/S0032-0633(01)00023-X[400].
- Russell, C.T. (1993). "Planetary Magnetospheres"[401] (PDF). *Reports on Progress in Physics*. **56** (6): 687–732. Bibcode: 1993RPPh...56..687R[402]. doi: 10.1088/0034-4885/56/6/001[403].
- Russell, C.T. (2001). "The dynamics of planetary magnetospheres". *Planetary and Space Science*. **49** (10–11): 1005–1030. Bibcode: 2001P&SS...49.1005R[404]. doi: 10.1016/S0032-0633(01)00017-4[405].
- Russell, C.T.; Khurana, K.K.; Arridge, C.S.; Dougherty, M.K. (2008). "The magnetospheres of Jupiter and Saturn and their lessons for the Earth"[406] (PDF). *Advances in Space Research*. **41** (8): 1310–18. Bibcode: 2008AdSpR..41.1310R[407]. doi: 10.1016/j.asr.2007.07.037[408].
- Santos-Costa, D.; Bourdarie, S.A. (2001). "Modeling the inner Jovian electron radiation belt including non-equatorial particles". *Planetary and Space Science*. **49** (3–4): 303–312. Bibcode: 2001P&SS...49..303S[409]. doi: 10.1016/S0032-0633(00)00151-3[410].
- Smith, E. J.; Davis, L. Jr.; et al. (1974). "The Planetary Magnetic Field and Magnetosphere of Jupiter: Pioneer 10". *Journal of Geophysical Research*. **79** (25): 3501–13. Bibcode: 1974JGR....79.3501S[411]. doi: 10.1029/JA079i025p03501[412].
- Troutman, P.A.; Bethke, K.; et al. (28 January 2003). "Revolutionary concepts for Human Outer Planet Exploration (HOPE)". *AIP Conference Proceedings*. **654**: 821–828. doi: 10.1063/1.1541373[413].
- Williams, D.J.; Mauk, B.; McEntire, R. W. (1998). "Properties of Ganymede's magnetosphere as revealed by energetic particle observations". *Journal of Geophysical Research*. **103** (A8): 17,523–534.

Bibcode: 1998JGR...10317523W[414]. doi: 10.1029/98JA01370[415].

• Wolverton, M. (2004). *The Depths of Space*. Joseph Henry Press. ISBN 978-0-309-09050-6.

• Zarka, P.; Kurth, W. S. (1998). "Auroral radio emissions at the outer planets: Observations and theory". *Journal of Geophysical Research*. **103** (E9): 20,159–194. Bibcode: 1998JGR...10320159Z[416]. doi: 10.1029/98JE01323[417].

• Zarka, P.; Kurth, W. S. (2005). "Radio wave emissions from the outer planets before Cassini". *Space Science Reviews*. **116** (1–2): 371–397. Bibcode: 2005SSRv..116..371Z[418]. doi: 10.1007/s11214-005-1962-2[419].

Further reading

Wikimedia Commons has media related to *Magnetosphere of Jupiter*.

<templatestyles src="Template:Refbegin/styles.css" />

• Carr, Thomas D.; Gulkis, Samuel (1969). "The magnetosphere of Jupiter". *Annual Review of Astronomy and Astrophysics*. **7** (1): 577–618. Bibcode: 1969ARA&A...7..577C[420]. doi: 10.1146/annurev.aa. 07.090169.003045[421].

• Edwards, T.M.; Bunce, E.J.; Cowley, S.W.H. (2001). "A note on the vector potential of Connerney et al.'s model of the equatorial current sheet in Jupiter's magnetosphere". *Planetary and Space Science*. **49** (10–11): 1115–23. Bibcode: 2001P&SS...49.1115E[422]. doi: 10.1016/ S0032-0633(00)00164-1[423].

• Gladstone, G.R.; Waite, J.H.; Grodent, D. (2002). "A pulsating auroral X-ray hot spot on Jupiter"[424]. *Nature*. **415** (6875): 1000–03. Bibcode: 2002Natur.415.1000G[425]. doi: 10.1038/4151000a[426]. PMID 11875561[427].

• Kivelson, Margaret G.; Khurana, Krishan K.; Walker, Raymond J. (2002). "Sheared magnetic field structure in Jupiter's dusk magnetosphere: Implications for return currents"[428] (PDF). *Journal of Geophysical Research*. **107** (A7): 1116. Bibcode: 2002JGRA..107.1116K[429]. doi: 10.1029/2001JA000251[430].

• Kivelson, M.G. (2005). "Transport and acceleration of plasma in the magnetospheres of Earth and Jupiter and expectations for Saturn"[431] (PDF). *Advances in Space Research*. **36** (11): 2077–89. Bibcode: 2005AdSpR..36.2077K[432]. doi: 10.1016/j.asr.2005.05.104[433].

- Kivelson, Margaret G.; Southwood, David J. (2003). "First evidence of IMF control of Jovian magnetospheric boundary locations: Cassini and Galileo magnetic field measurements compared"[434] (PDF). *Planetary and Space Science*. **51** (A7): 891–98. Bibcode: 2003P&SS...51..891K[435]. doi: 10.1016/S0032-0633(03)00075-8[436].

- McComas, D.J.; Allegrini, F.; Bagenal, F.; et al. (2007). "Diverse Plasma Populations and Structures in Jupiter's Magnetotail". *Science*. **318** (5848): 217–20. Bibcode: 2007Sci...318..217M[437]. doi: 10.1126/science.1147393[438]. PMID 17932282[439].

- Maclennan, G.G.; Maclennan, L.J.; Lagg, Andreas (2001). "Hot plasma heavy ion abundance in the inner Jovian magnetosphere (<10 Rj)". *Planetary and Space Science*. **49** (3–4): 275–82. Bibcode: 2001P&SS...49..275M[440]. doi: 10.1016/S0032-0633(00)00148-3[441].

- Russell, C.T.; Yu, Z.J.; Kivelson, M.G. (2001). "The rotation period of Jupiter"[442] (PDF). *Geophysical Research Letters*. **28** (10): 1911–12. Bibcode: 2001GeoRL..28.1911R[443]. doi: 10.1029/2001GL012917[444].

- Zarka, Philippe; Queinnec, Julien; Crary, Frank J. (2001). "Low-frequency limit of Jovian radio emissions and implications on source locations and Io plasma wake". *Planetary and Space Science*. **49** (10–11): 1137–49. Bibcode: 2001P&SS...49.1137Z[445]. doi: 10.1016/S0032-0633(01)00021-6[446].

<indicator name="featured-star"> ✦ </indicator>

Exploration

Exploration of Jupiter

Summary of missions to the outer Solar System

System Spacecraft	Jupiter	Saturn	Uranus	Neptune	Pluto
Pioneer 10	**1973** flyby				
Pioneer 11	**1974** flyby	**1979** flyby			
Voyager 1	**1979** flyby	**1980** flyby			
Voyager 2	**1979** flyby	**1981** flyby	**1986** flyby	**1989** flyby	
Galileo	**1995–2003** orbiter; **1995, 2003** atmospheric				
Ulysses	**1992, 2004** gravity assist				
Cassini–Huygens	**2000** gravity assist	**2004–2017** orbiter; **2005** Titan lander			
New Horizons	**2007** gravity assist				**2015** flyby
Juno	**2016–** orbiter				
Jupiter Icy Moons Explorer	**2022–** Planned orbiter				
Europa Clipper	**2025–** Planned orbiter				

The **exploration of Jupiter** has been conducted via close observations by automated spacecraft. It began with the arrival of *Pioneer 10* into the Jovian system in 1973, and, as of 2016[447], has continued with eight further spacecraft missions. All of these missions were undertaken by the National Aeronautics and Space Administration (NASA), and all but two have been flybys that

Figure 57: *Jupiter as seen by the space probe Cassini*

take detailed observations without the probe landing or entering orbit. These probes make Jupiter the most visited of the Solar System's outer planets as all missions to the outer Solar System have used Jupiter flybys to reduce fuel requirements and travel time. On 5 July 2016, spacecraft *Juno* arrived and entered the planet's orbit—the second craft ever to do so. Sending a craft to Jupiter entails many technical difficulties, especially due to the probes' large fuel requirements and the effects of the planet's harsh radiation environment.

The first spacecraft to visit Jupiter was *Pioneer 10* in 1973, followed a year later by *Pioneer 11*. Aside from taking the first close-up pictures of the planet, the probes discovered its magnetosphere and its largely fluid interior. The *Voyager 1* and *Voyager 2* probes visited the planet in 1979, and studied its moons and the ring system, discovering the volcanic activity of Io and the presence of water ice on the surface of Europa. *Ulysses* further studied Jupiter's magnetosphere in 1992 and then again in 2000. The *Cassini* probe approached the planet in 2000 and took very detailed images of its atmosphere. The *New Horizons* spacecraft passed by Jupiter in 2007 and made improved measurements of its and its satellites' parameters.

The *Galileo* spacecraft was the first to have entered orbit around Jupiter, arriving in 1995 and studying the planet until 2003. During this period *Galileo* gathered a large amount of information about the Jovian system, making close approaches to all of the four large Galilean moons and finding evidence for thin atmospheres on three of them, as well as the possibility of liquid water beneath their surfaces. It also discovered a magnetic field around Ganymede. As it approached Jupiter, it also witnessed the impact of Comet Shoemaker–Levy 9. In December 1995, it sent an atmospheric probe into the Jovian atmosphere, so far the only craft to do so.

In July 2016, the *Juno* spacecraft, launched in 2011, completed its orbital insertion maneuver successfully, and is now in orbit around Jupiter with its science programme ongoing.

The European Space Agency selected the L1-class JUICE mission in 2012 as part of its Cosmic Vision programme to explore three of Jupiter's Galilean moons, with a possible Ganymede lander provided by Roscosmos. JUICE is proposed to be launched in 2022. ISRO will launch the first Indian mission to Jupiter in 2022 or 2023 through Geosynchronous Satellite Launch Vehicle Mark III.

Technical requirements

Flights from Earth to other planets in the Solar System have a high energy cost. It requires almost the same amount of energy for a spacecraft to reach Jupiter from Earth's orbit as it does to lift it into orbit in the first place. In astrodynamics, this energy expenditure is defined by the net change in the spacecraft's velocity, or delta-v. The energy needed to reach Jupiter from an Earth orbit requires a delta-v of about 9 km/s, compared to the 9.0–9.5 km/s to reach a low Earth orbit from the ground. Gravity assists through planetary flybys (such as by Earth or Venus) can be used to reduce the energetic requirement (i.e. the fuel) at launch, at the cost of a significantly longer flight duration to reach a target such as Jupiter when compared to the direct trajectory.[448] Ion thrusters capable of a delta-v of more than 10 kilometers/s were used on the Dawn spacecraft. This is more than enough delta-v to do a Jupiter fly-by mission from a solar orbit of the same radius as that of Earth without gravity assist.[449]

A major problem in sending space probes to Jupiter is that the planet has no solid surface on which to land, as there is a smooth transition between the planet's atmosphere and its fluid interior. Any probes descending into the atmosphere are eventually crushed by the immense pressures within Jupiter.

Another major issue is the amount of radiation to which a space probe is subjected, due to the harsh charged-particle environment around Jupiter (for a detailed explanation see Magnetosphere of Jupiter). For example, when *Pioneer 11* made its closest approach to the planet, the level of radiation was ten times more powerful than *Pioneer*'s designers had predicted, leading to fears that the probes would not survive. With a few minor glitches, the probe managed to pass through the radiation belts, but it lost most of the images of the moon Io, as the radiation had caused *Pioneer*'s imaging photo polarimeter to receive false commands. The subsequent and far more technologically advanced *Voyager* spacecraft had to be redesigned to cope with the radiation levels. Over the eight years the *Galileo* spacecraft orbited the planet, the probe's radiation dose far exceeded its design specifications, and its systems failed on several occasions. The spacecraft's gyroscopes often exhibited increased errors, and electrical arcs sometimes occurred between its rotating and non-rotating parts, causing it to enter safe mode, which led to total loss of the data from the 16th, 18th and 33rd orbits. The radiation also caused phase shifts in *Galileo*'s ultra-stable quartz oscillator.

Flyby missions

<templatestyles src="Multiple_image/styles.css" />

South pole (*Cassini*; 2000)

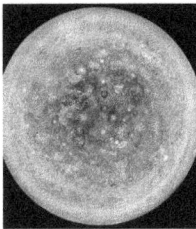

South pole (*Juno*; 2017)

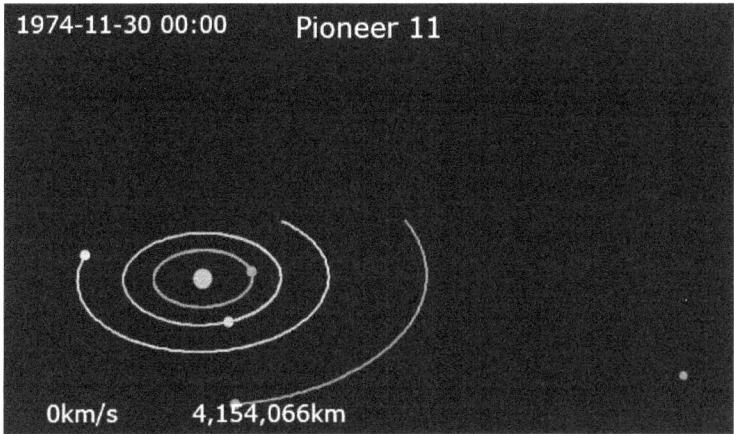

Figure 58: *Animation of Pioneer 11's trajectory around Jupiter from 30 November 1974 to 5 December 1974*

Pioneer 11 · Jupiter · Io · Europa · Ganymede · Callisto

Figure 59: *Pioneer 10 was the first spacecraft to visit Jupiter.*

Figure 60: *Time-lapse sequence from the approach of Voyager 1 to Jupiter*

Pioneer program (1973 and 1974)

The first spacecraft to explore Jupiter was *Pioneer 10*, which flew past the planet in December 1973, followed by *Pioneer 11* twelve months later. *Pioneer 10* obtained the first-ever close-up images of Jupiter and its Galilean moons; the spacecraft studied the planet's atmosphere, detected its magnetic field, observed its radiation belts and determined that Jupiter is mainly fluid. *Pioneer 11* made its closest approach, within some 34,000 km of Jupiter's cloud tops, on December 4, 1974. It obtained dramatic images of the Great Red Spot, made the first observation of Jupiter's immense polar regions, and determined the mass of Jupiter's moon Callisto. The information gathered by these two spacecraft helped astronomers and engineers improve the design of future probes to cope more effectively with the environment around the giant planet.

Voyager program (1979)

Voyager 1 began photographing Jupiter in January 1979 and made its closest approach on March 5, 1979, at a distance of 349,000 km from Jupiter's center. This close approach allowed for greater image resolution, though the flyby's short duration meant that most observations of Jupiter's moons, rings, magnetic field, and radiation environment were made in the 48-hour period bracketing the approach, even though *Voyager 1* continued photographing the planet until April. It was soon followed by *Voyager 2*, which made its closest approach on July 9, 1979, 576,000 km away from the planet's cloud tops. The probe discovered Jupiter's ring, observed intricate vortices in its atmosphere, observed active volcanoes on Io, a process analogous to plate tectonics on Ganymede, and numerous craters on Callisto.

The *Voyager* missions vastly improved our understanding of the Galilean moons, and also discovered Jupiter's rings. They also took the first close-up images of the planet's atmosphere, revealing the Great Red Spot as a complex storm moving in a counter-clockwise direction. Other smaller storms and eddies were found throughout the banded clouds (see animation on the right). Two new, small satellites, Adrastea and Metis, were discovered orbiting just outside the ring, making them the first of Jupiter's moons to be identified by a spacecraft.[450] A third new satellite, Thebe, was discovered between the orbits of Amalthea and Io.

The discovery of volcanic activity on the moon Io was the greatest unexpected finding of the mission, as it was the first time an active volcano was observed on a celestial body other than Earth. Together, the *Voyagers* recorded the eruption of nine volcanoes on Io, as well as evidence for other eruptions occurring between the Voyager encounters.

Europa displayed a large number of intersecting linear features in the low-resolution photos from *Voyager 1*. At first, scientists believed the features might be deep cracks, caused by crustal rifting or tectonic processes. The high-resolution photos from *Voyager 2*, taken closer to Jupiter, left scientists puzzled as the features in these photos were almost entirely lacking in topographic relief. This led many to suggest that these cracks might be similar to ice floes on Earth, and that Europa might have a liquid water interior. Europa may be internally active due to tidal heating at a level about one-tenth that of Io, and as a result, the moon is thought to have a thin crust less than 30 kilometers (19 mi) thick of water ice, possibly floating on a 50-kilometers-deep (30 mile) ocean.

Ulysses (1992)

On February 8, 1992, the *Ulysses* solar probe flew past Jupiter's north pole at a distance of 451,000 km. This swing-by maneuver was required for *Ulysses* to attain a very high-inclination orbit around the Sun, increasing its inclination to the ecliptic to 80.2 degrees. The giant planet's gravity bent the spacecraft's flightpath downward and away from the ecliptic plane, placing it into a final orbit around the Sun's north and south poles. The size and shape of the probe's orbit were adjusted to a much smaller degree, so that its aphelion remained at approximately 5 AU (Jupiter's distance from the Sun), while its perihelion lay somewhat beyond 1 AU (Earth's distance from the Sun). During its Jupiter encounter, the probe made measurements of the planet's magnetosphere. Since the probe had no cameras, no images were taken. In February 2004, the probe arrived again at the vicinity of Jupiter. This time the distance from the planet was much greater—about 120 million km (0.8 AU)—but it made further observations of Jupiter.

Cassini (2000)

In 2000, the Cassini probe, en route to Saturn, flew by Jupiter and provided some of the highest-resolution images ever taken of the planet. It made its closest approach on December 30, 2000, and made many scientific measurements. About 26,000 images of Jupiter were taken during the months-long flyby. It produced the most detailed global color portrait of Jupiter yet, in which the smallest visible features are approximately 60 km (37 mi) across.

A major finding of the flyby, announced on March 5, 2003, was of Jupiter's atmospheric circulation. Dark belts alternate with light zones in the atmosphere, and the zones, with their pale clouds, had previously been considered by scientists to be areas of upwelling air, partly because on Earth clouds tend to be formed by rising air. Analysis of Cassini imagery showed that the dark belts contain individual storm cells of upwelling bright-white clouds, too small to see from Earth. Anthony Del Genio of NASA's Goddard Institute for Space Studies said that "the belts must be the areas of net-rising atmospheric motion on Jupiter, [so] the net motion in the zones has to be sinking".

Other atmospheric observations included a swirling dark oval of high atmospheric-haze, about the size of the Great Red Spot, near Jupiter's north pole. Infrared imagery revealed aspects of circulation near the poles, with bands of globe-encircling winds, and adjacent bands moving in opposite directions. The same announcement also discussed the nature of Jupiter's rings. Light scattering by particles in the rings showed the particles were irregularly shaped (rather than spherical) and likely originated as ejecta from micrometeorite impacts on Jupiter's moons, probably on Metis and Adrastea. On December 19, 2000, the Cassini spacecraft captured a very-low-resolution image of the moon Himalia, but it was too distant to show any surface details.

New Horizons (2007)

The New Horizons probe, en route to Pluto, flew by Jupiter for a gravity assist and was the first probe launched directly towards Jupiter since the Ulysses in 1990. Its Long Range Reconnaissance Imager (LORRI) took its first photographs of Jupiter on September 4, 2006. The spacecraft began further study

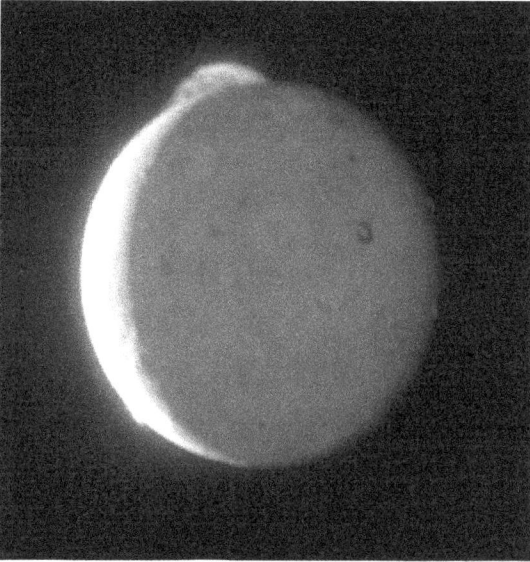

Figure 61: *Video of volcanic plumes on Io, as recorded by New Horizons in 2008*

of the Jovian system in December 2006, and made its closest approach on February 28, 2007.

Although close to Jupiter, *New Horizons'* instruments made refined measurements of the orbits of Jupiter's inner moons, particularly Amalthea. The probe's cameras measured volcanoes on Io, studied all four Galilean moons in detail, and made long-distance studies of the outer moons Himalia and Elara. The craft also studied Jupiter's Little Red Spot and the planet's magnetosphere and tenuous ring system.

On March 19, 2007 the Command and Data Handling computer experienced an uncorrectable memory error and rebooted itself, causing the spacecraft to go into safe mode. The craft fully recovered within two days, with some data loss on Jupiter's magnetotail. No other data loss events were associated with the encounter. Due to the immense size of the Jupiter system and the relative closeness of the Jovian system to Earth in comparison to the closeness of Pluto to Earth, *New Horizons* sent back more data to Earth from the Jupiter encounter than the Pluto encounter.

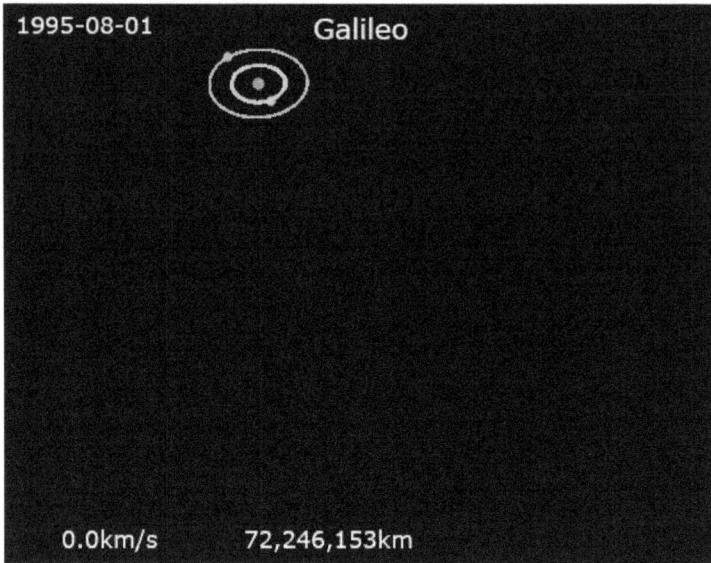

Figure 62: *Animation of Galileo's trajectory around Jupiter from 1 August 1995 to 30 September 2003*
Galileo · Jupiter · Ganymede · Callisto

Orbiter missions

Galileo (1995–2003)

The first spacecraft to orbit Jupiter was the *Galileo* orbiter, which went into orbit around Jupiter on December 7, 1995. It orbited the planet for over seven years, making 35 orbits before it was destroyed during a controlled impact with Jupiter on September 21, 2003. During this period, it gathered a large amount of information about the Jovian system; the amount of information was not as great as intended because the deployment of its high-gain radio transmitting antenna failed. The major events during the eight-year study included multiple flybys of all of the Galilean moons, as well as Amalthea (the first probe to do so). It also witnessed the impact of Comet Shoemaker–Levy 9 as it approached Jupiter in 1994 and the sending of an atmospheric probe into the Jovian atmosphere in December 1995.

Cameras on the *Galileo* spacecraft observed fragments of Comet Shoemaker–Levy 9 between 16 and 22 July 1994 as they collided with Jupiter's southern hemisphere at a speed of approximately 60 kilometres per second. This was the first direct observation of an extraterrestrial collision of solar system objects. While the impacts took place on the side of Jupiter hidden from

Figure 63: *A sequence of Galileo images taken several seconds apart shows the appearance of the fireball appearing on the dark side of Jupiter from one of the fragments of Comet Shoemaker–Levy 9 hitting the planet.*

Earth, *Galileo*, then at a distance of 1.6 AU from the planet, was able to see the impacts as they occurred. Its instruments detected a fireball that reached a peak temperature of about 24,000 K, compared to the typical Jovian cloud-top temperature of about 130 K (–143 °C), with the plume from the fireball reaching a height of over 3,000 km.

An atmospheric probe was released from the spacecraft in July 1995, entering the planet's atmosphere on December 7, 1995. After a high-g descent into the Jovian atmosphere, the probe discarded the remains of its heat shield, and it parachuted through 150 km of the atmosphere, collecting data for 57.6 minutes, before being crushed by the pressure and temperature to which it was subjected (about 22 times Earth normal, at a temperature of 153 °C). It would have melted thereafter, and possibly vaporized. The *Galileo* orbiter itself experienced a more rapid version of the same fate when it was deliberately steered into the planet on September 21, 2003 at a speed of over 50 km/s, in order to avoid any possibility of it crashing into and contaminating Europa.

Major scientific results of the *Galileo* mission include:

- the first observation of ammonia clouds in another planet's atmosphere—the atmosphere creates ammonia ice particles from material coming up from lower depths;
- confirmation of extensive volcanic activity on Io—which is 100 times greater than that found on Earth; the heat and frequency of eruptions are reminiscent of early Earth;
- observation of complex plasma interactions in Io's atmosphere which create immense electrical currents that couple to Jupiter's atmosphere;
- providing evidence for supporting the theory that liquid oceans exist under Europa's icy surface;

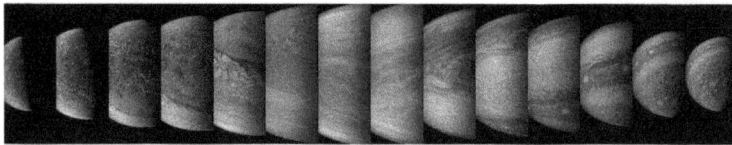

Figure 64:
Jupiter – perijove pass as viewed by JunoCam.

- first detection of a substantial magnetic field around a satellite (Ganymede);
- magnetic data evidence suggesting that Europa, Ganymede and Callisto have a liquid-saltwater layer under the visible surface;
- evidence for a thin atmospheric layer on Europa, Ganymede, and Callisto known as a 'surface-bound exosphere';
- understanding of the formation of the rings of Jupiter (by dust kicked up as interplanetary meteoroids which smash into the planet's four small inner moons) and observation of two outer rings and the possibility of a separate ring along Amalthea's orbit;
- identification of the global structure and dynamics of a giant planet's magnetosphere.

On December 11, 2013, NASA reported, based on results from the Galileo mission, the detection of "clay-like minerals" (specifically, phyllosilicates), often associated with organic materials, on the icy crust of Europa, moon of Jupiter. The presence of the minerals may have been the result of a collision with an asteroid or comet according to the scientists.

Juno (2016)

NASA launched *Juno* on August 5, 2011 to study Jupiter in detail. It entered a polar orbit of Jupiter on July 5, 2016. The spacecraft is studying the planet's composition, gravity field, magnetic field, and polar magnetosphere. *Juno* is also searching for clues about how Jupiter formed, including whether the planet has a rocky core, the amount of water present within the deep atmosphere, and how the mass is distributed within the planet. *Juno* also studies Jupiter's deep winds,[451] which can reach speeds of 600 km/h.

Jupiter Icy Moon Explorer (2022)

ESA's Jupiter Icy Moon Explorer (JUICE) has been selected as part of ESA's Cosmic Vision science program. It is expected to launch in 2022 and, after a series of flybys in the inner Solar System, arrive in 2030. In 2012, the European Space Agency's selected the *JUpiter ICy moon Explorer* (JUICE) as its

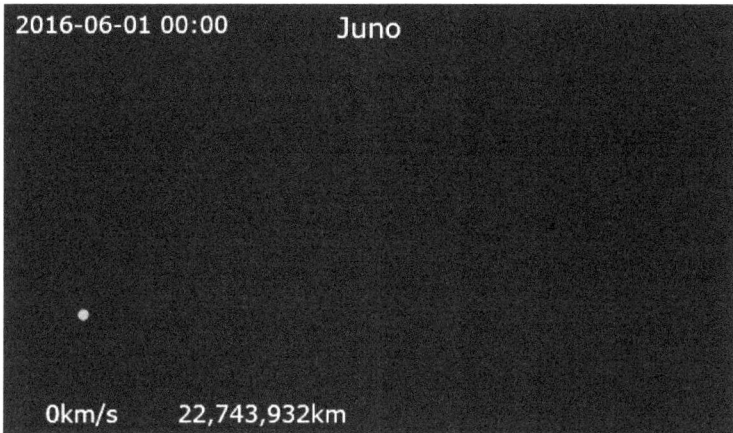

Figure 65: *Animation of Juno's trajectory around
Jupiter from 1 June 2016 to 31 July 2021*
Juno · Jupiter

first Large mission, replacing its contribution to EJSM, the *Jupiter Ganymede
Orbiter* (JGO). The partnership for the Europa Jupiter System Mission has
since ended, but NASA will continue to contribute the European mission with
hardware and an instrument.

Proposed missions

The *Europa Clipper* is a mission proposed to NASA to focus on studying
Jupiter's moon Europa. In March 2013, funds were authorized for "pre-
formulation and/or formulation activities for a mission that meets the science
goals outlined for the Jupiter Europa mission in the most recent planetary
decadal survey". The proposed mission would be set to launch in the early
2020s and reach Europa after a 6.5 year cruise. The spacecraft would fly by
the moon 32 times to minimize radiation damage.

Canceled missions

Because of the possibility of subsurface liquid oceans on Jupiter's moons Europa, Ganymede and Callisto, there has been great interest in studying the icy
moons in detail. Funding difficulties have delayed progress. The *Europa Orbiter* was a planned NASA mission to Europa, which was canceled in 2002.
Its main objectives included determining the presence or absence of a subsurface ocean and identifying candidate sites for future lander missions. NASA's

JIMO (*Jupiter Icy Moons Orbiter*), which was canceled in 2005, and a European *Jovian Europa Orbiter* mission were also studied, but were superseded by the *Europa Jupiter System Mission*.

The *Europa Jupiter System Mission* (EJSM) was a joint NASA/ESA proposal for exploration of Jupiter and its moons. In February 2009 it was announced that both space agencies had given this mission priority ahead of the *Titan Saturn System Mission*. The proposal included a launch date of around 2020 and consists of the NASA-led *Jupiter Europa Orbiter*, and the ESA-led *Jupiter Ganymede Orbiter*. ESA's contribution had encountered funding competition from other ESA projects. However, the *Jupiter Europa Orbiter* (JEO), NASA's contribution, was considered by the Planetary Decadal Survey to be too expensive. The survey supported a cheaper alternative to JEO.[452]

Human exploration

While scientists require further evidence to determine the extent of a rocky core on Jupiter, its Galilean moons provide the potential opportunity for future human exploration.

Particular targets are Europa, due to its potential for life, and Callisto, due to its relatively low radiation dose.[453] In 2003, NASA proposed a program called Human Outer Planets Exploration (HOPE) that involved sending astronauts to explore the Galilean moons. NASA has projected a possible attempt some time in the 2040s. In the Vision for Space Exploration policy announced in January 2004, NASA discussed missions beyond Mars, mentioning that a "human research presence" may be desirable on Jupiter's moons. Before the JIMO mission was cancelled, NASA administrator Sean O'Keefe stated that "human explorers will follow."

Potential for colonization

NASA has speculated on the feasibility of mining the atmospheres of the outer planets, particularly for helium-3, an isotope of helium that is rare on Earth and could have a very high value per unit mass as thermonuclear fuel.[454,455] Factories stationed in orbit could mine the gas and deliver it to visiting craft. However, the Jovian system in general poses particular disadvantages for colonization because of the severe radiation conditions prevailing in Jupiter's magnetosphere and the planet's particularly deep gravitational well. Jupiter would deliver about 36 Sv (3600 rem) per day to unshielded colonists at Io and about 5.4 Sv (540 rems) per day to unshielded colonists at Europa, which is a decisive aspect due to the fact that already an exposure to about 0.75 Sv over a period of a few days is enough to cause radiation poisoning, and about 5 Sv over a few days is fatal.[456]

Jovian radiation

Moon	rem/day
Io	3600
Europa	540
Ganymede	8
Callisto	0.01
Earth (Max)	**0.07**
Earth (Avg)	**0.0007**

Ganymede is the Solar System's largest moon and the Solar System's only known moon with a magnetosphere, but this does not shield it from cosmic radiation to a noteworthy degree, because it is overshadowed by Jupiter's magnetic field. Ganymede receives about 0.08 Sv (8 rem) of radiation per day. Callisto is farther from Jupiter's strong radiation belt and subject to only 0.0001 Sv (0.01 rem) a day. For comparison, the average amount of radiation taken on Earth by a living organism is about 0.0024 Sv per year; the highest natural radiation levels on Earth are recorded around Ramsar hot springs at about 0.26 Sv per year.

One of the main targets chosen by the HOPE study was Callisto. The possibility of building a surface base on Callisto was proposed, because of the low radiation levels at its distance from Jupiter and its geological stability. Callisto is the only Galilean satellite for which human exploration is feasible. The levels of ionizing radiation on Io, Europa, and Ganymede are hostile to human life, and adequate protective measures have yet to be devised.

It could be possible to build a surface base that would produce fuel for further exploration of the Solar System. In 1997, the Artemis Project designed a plan to colonize Europa. According to this plan, explorers would drill down into the Europan ice crust, entering the postulated subsurface ocean, where they would inhabit artificial air pockets.

External links

- Chronology of Lunar and Planetary Exploration[457]
- NASA missions to Jupiter[458]

<indicator name="good-star"> ⊕ </indicator>

Moons of Jupiter

Moons of Jupiter

There are 79 known **moons of Jupiter**. This gives Jupiter the largest number of moons with reasonably stable orbits of any planet in the Solar System. The most massive of the moons are the four Galilean moons, which were independently discovered in 1610 by Galileo Galilei and Simon Marius and were the first objects found to orbit a body that was neither Earth nor the Sun. From the end of the 19th century, dozens of much smaller Jovian moons have been discovered and have received the names of lovers or daughters of the Roman god Jupiter or his Greek equivalent Zeus. The Galilean moons are by far the largest and most massive objects to orbit Jupiter, with the remaining 75 known moons and the rings together comprising just 0.003% of the total orbiting mass.

Of Jupiter's moons, eight are regular satellites with prograde and nearly circular orbits that are not greatly inclined with respect to Jupiter's equatorial plane. The Galilean satellites are nearly spherical in shape due to their planetary mass, and so would be considered at least dwarf planets if they were in direct orbit around the Sun. The other four regular satellites are much smaller and closer to Jupiter; these serve as sources of the dust that makes up Jupiter's rings. The remainder of Jupiter's moons are irregular satellites whose prograde and retrograde orbits are much farther from Jupiter and have high inclinations and eccentricities. These moons were probably captured by Jupiter from solar orbits. Twenty-eight of the irregular satellites have not yet been officially named.

Characteristics

The physical and orbital characteristics of the moons vary widely. The four Galileans are all over 3,100 kilometres (1,900 mi) in diameter; the largest Galilean, Ganymede, is the ninth largest object in the Solar System, after the Sun and seven of the planets, Ganymede being larger than Mercury. All other

Figure 66: *A montage of Jupiter and its four largest moons (distance and sizes not to scale)*

Figure 67: *The orbit and motion of the Galilean moons around Jupiter, as captured by JunoCam aboard the Juno spacecraft.*

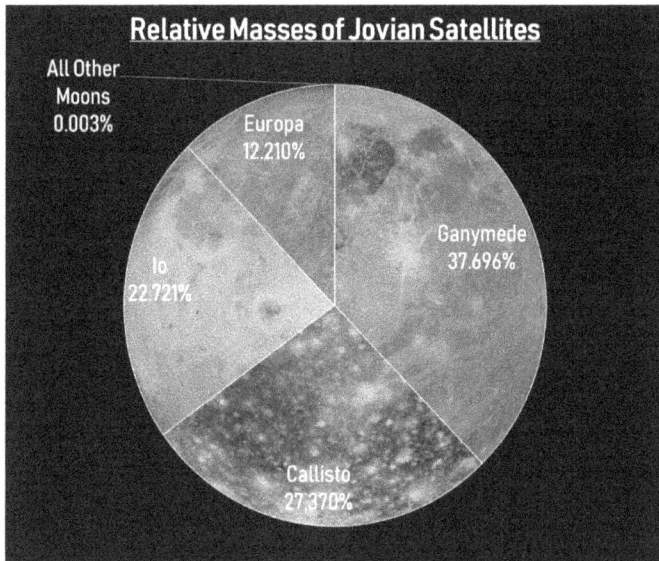

Figure 68: *The relative masses of the Jovian moons. Those smaller than Europa are not visible at this scale, and combined would only be visible at 100× magnification.*

Jovian moons are less than 250 kilometres (160 mi) in diameter, with most barely exceeding 5 kilometres (3.1 mi).[459] Their orbital shapes range from nearly perfectly circular to highly eccentric and inclined, and many revolve in the direction opposite to Jupiter's spin (retrograde motion). Orbital periods range from seven hours (taking less time than Jupiter does to spin around its axis), to some three thousand times more (almost three Earth years).

Origin and evolution

Jupiter's regular satellites are believed to have formed from a circumplanetary disk, a ring of accreting gas and solid debris analogous to a protoplanetary disk. They may be the remnants of a score of Galilean-mass satellites that formed early in Jupiter's history.

Simulations suggest that, while the disk had a relatively high mass at any given moment, over time a substantial fraction (several tenths of a percent) of the mass of Jupiter captured from the solar nebula was passed through it. However, only 2% of the proto-disk mass of Jupiter is required to explain the existing satellites. Thus there may have been several generations of Galilean-mass

Figure 69: *The Galilean moons. From left to right, in order of increasing distance from Jupiter: Io; Europa; Ganymede; Callisto.*

satellites in Jupiter's early history. Each generation of moons might have spiraled into Jupiter, because of drag from the disk, with new moons then forming from the new debris captured from the solar nebula. By the time the present (possibly fifth) generation formed, the disk had thinned so that it no longer greatly interfered with the moons' orbits. The current Galilean moons were still affected, falling into and being partially protected by an orbital resonance with each other, which still exists for Io, Europa, and Ganymede. Ganymede's larger mass means that it would have migrated inward at a faster rate than Europa or Io.

The outer, irregular moons are thought to have originated from captured asteroids, whereas the protolunar disk was still massive enough to absorb much of their momentum and thus capture them into orbit. Many are believed to have broken up by mechanical stresses during capture, or afterward by collisions with other small bodies, producing the moons we see today.

Discovery

Some scholars propose that the earliest record of a Jovian moon (Ganymede or Callisto) is a note by Chinese astronomer Gan De of an observation around 364 BC.

However, the first certain observations of Jupiter's satellites were those of Galileo Galilei in 1609. By January 1610, he had sighted the four massive Galilean moons with his $30\times$ magnification telescope, and he published his results in March 1610.

Simon Marius had independently discovered the moons one day after Galileo, although he did not publish his book on the subject until 1614. Even so, the names Marius assigned are used today: Ganymede; Callisto; Io; and Europa. No additional satellites were discovered until E. E. Barnard observed Amalthea in 1892.

Figure 70: *Jupiter and the Galilean moons through a 25 cm (10 in) Meade LX200 telescope.*

With the aid of telescopic photography, further discoveries followed quickly over the course of the 20th century. Himalia was discovered in 1904, Elara in 1905, Pasiphae in 1908, Sinope in 1914, Lysithea and Carme in 1938, Ananke in 1951, and Leda in 1974.

By the time that the Voyager space probes reached Jupiter, around 1979, 13 moons had been discovered, not including Themisto, which had been observed in 1975, but was lost until 2000 due to insufficient initial observation data. The Voyager spacecraft discovered an additional three inner moons in 1979: Metis; Adrastea; and Thebe.

No additional moons were discovered for two decades, generally during the 1980s and 1990s, but between October 1999 and February 2003, researchers found another 34 moons using sensitive ground-based detectors. These are tiny moons, in long, eccentric, generally retrograde orbits, and averaging 3 km (1.9 mi) in diameter, with the largest being just 9 km (5.6 mi) across. All of these moons are thought to have been captured asteroidal or perhaps comet bodies, possibly fragmented into several pieces.

By 2015, a total of 15 additional moons were discovered. Two more were discovered in 2016 by the team led by Scott S. Sheppard at the Carnegie Institution for Science, bringing the total to 69. On 17 July 2018, the International Astronomical Union confirmed that Sheppard's team discovered ten more moons around Jupiter, bringing the total number to 79. Among these is S/2016 J 2, which has a prograde orbit, but crosses paths with several moons

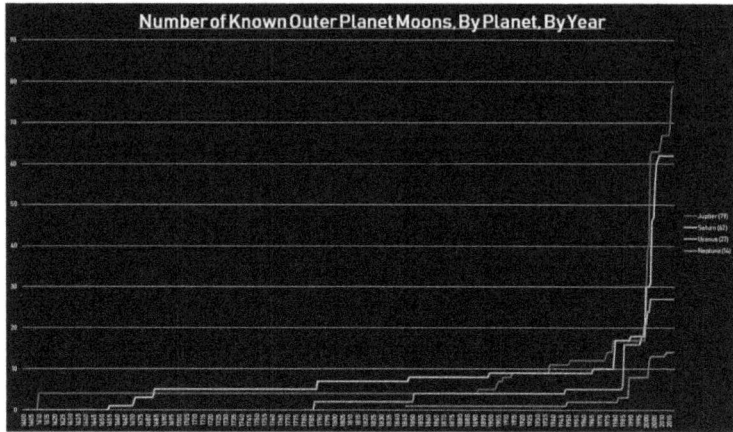

Figure 71: *The number of moons known for each of the four outer planets up to July 2018. Jupiter currently has 79 known satellites.*

that have retrograde orbits, making an eventual collision—at some point on a billions of years timescale—likely.

Additional tiny moons likely exist but remain undiscovered, as they are very difficult for astronomers to detect.

Naming

The Galilean moons of Jupiter (Io, Europa, Ganymede, and Callisto) were named by Simon Marius soon after their discovery in 1610. However, these names fell out of favor until the 20th century. The astronomical literature instead simply referred to "Jupiter I", "Jupiter II", etc., or "the first satellite of Jupiter", "Jupiter's second satellite", and so on. The names Io, Europa, Ganymede, and Callisto became popular in the 20th century, whereas the rest of the moons remained unnamed and were usually numbered in Roman numerals V (5) to XII (12). Jupiter V was discovered in 1892 and given the name *Amalthea* by a popular though unofficial convention, a name first used by French astronomer Camille Flammarion.

The other moons were simply labeled by their Roman numeral (e.g. Jupiter IX) in the majority of astronomical literature until the 1970s. In 1975, the International Astronomical Union's (IAU) Task Group for Outer Solar System Nomenclature granted names to satellites V–XIII, and provided for a formal naming process for future satellites still to be discovered. The practice was to name newly discovered moons of Jupiter after lovers and favorites of the

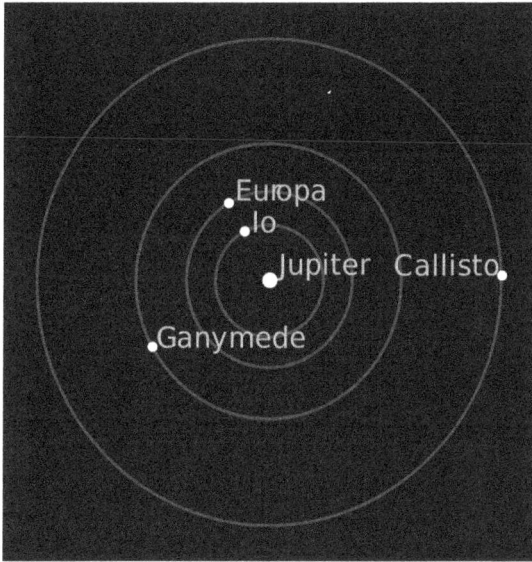

Figure 72: *The Galilean moons and their orbits around Jupiter.*

god Jupiter (Zeus) and, since 2004, also after their descendants.[460] All of Jupiter's satellites from XXXIV (Euporie) onward except LIII (Dia) are named after daughters of Jupiter or Zeus. Names ending with "a" or "o" are used for prograde irregular satellites (the latter for highly inclined satellites), and names ending with "e" are used for retrograde irregulars. The most recently confirmed moons Jupiter LI through LX (with the exception of Jupiter LIII Dia) have not received names.

Some asteroids share the same names as moons of Jupiter: 9 Metis, 38 Leda, 52 Europa, 85 Io, 113 Amalthea, 239 Adrastea. Two more asteroids previously shared the names of Jovian moons until spelling differences were made permanent by the IAU: Ganymede and asteroid 1036 Ganymed; and Callisto and asteroid 204 Kallisto.

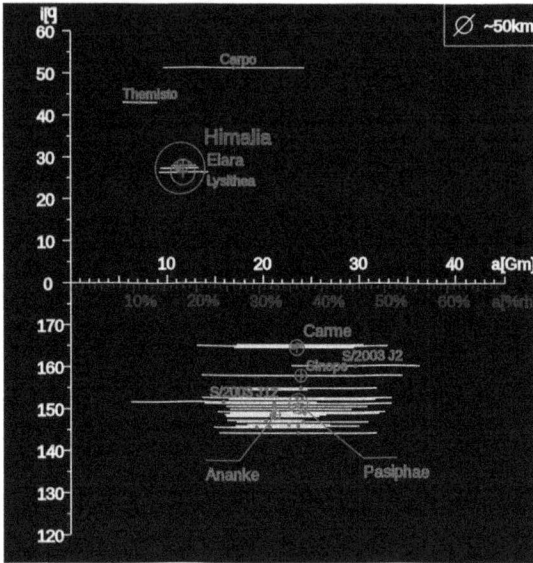

Figure 73: *The orbits of Jupiter's irregular satellites, and how they cluster into groups: by semi-major axis (the horizontal axis in Gm); by orbital inclination (the vertical axis); and orbital eccentricity (the yellow lines). The relative sizes are indicated by the circles.*

Groups

Regular satellites

These have prograde and nearly circular orbits of low inclination and are split into two groups:

- **Inner satellites** or **Amalthea group**: Metis, Adrastea, Amalthea, and Thebe. These orbit very close to Jupiter; the innermost two orbit in less than a Jovian day. The latter two are respectively the fifth and seventh largest moons in the Jovian system. Observations suggest that at least the largest member, Amalthea, did not form on its present orbit, but farther from the planet, or that it is a captured Solar System body. These moons, along with a number of as-yet-unseen inner moonlets, replenish and maintain Jupiter's faint ring system. Metis and Adrastea help to maintain Jupiter's main ring, whereas Amalthea and Thebe each maintain their own faint outer rings.

- **Main group** or **Galilean moons**: Io, Europa, Ganymede and Callisto. They are some of the largest objects in the Solar System outside the

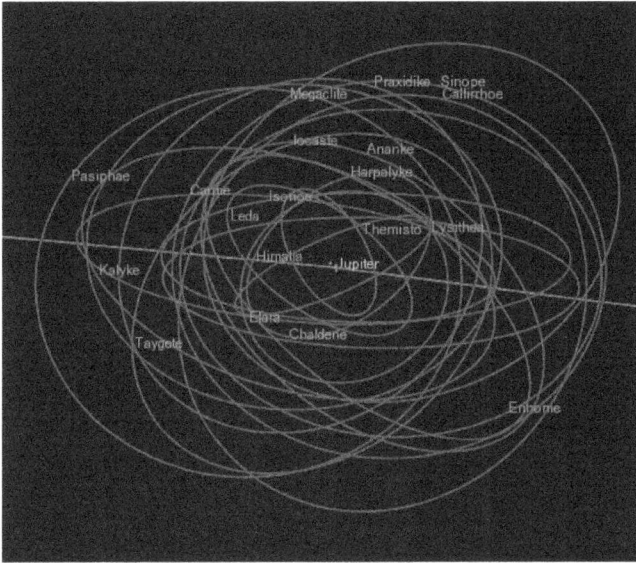

Figure 74: *Jupiter's outer moons and their highly inclined orbits*

Sun and the eight planets in terms of mass and are larger than any
known dwarf planet. Ganymede exceeds even the planet Mercury in
diameter. They are respectively the fourth-, sixth-, first-, and third-
largest natural satellites in the Solar System, containing approximately
99.997% of the total mass in orbit around Jupiter, while Jupiter is
almost 5,000 times more massive than the Galilean moons.[461] The
inner moons are in a 1:2:4 orbital resonance. Models suggest that
they formed by slow accretion in the low-density Jovian subneb-
ula—a disc of the gas and dust that existed around Jupiter after its
formation—which lasted up to 10 million years in the case of Callisto.
Several are suspected of having subsurface oceans.

Irregular satellites

The irregular satellites are substantially smaller objects with more distant and
eccentric orbits. They form families with shared similarities in orbit (semi-
major axis, inclination, eccentricity) and composition; it is believed that these
are at least partially collisional families that were created when larger (but
still small) parent bodies were shattered by impacts from asteroids captured
by Jupiter's gravitational field. These families bear the names of their largest
members. The identification of satellite families is tentative, but the following
are typically listed:

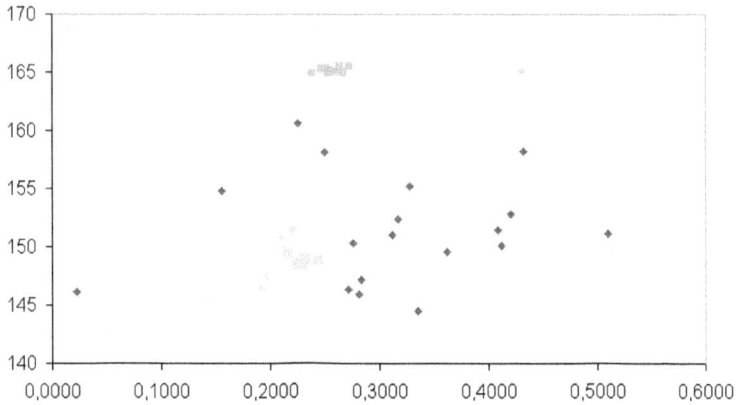

Figure 75: *Retrograde satellites: inclinations (°) vs. eccentricities, with Carme's (orange) and Ananke's (yellow) groups identified*

- Prograde satellites:
  - Themisto is the innermost irregular moon and is not part of a known family.
  - The **Himalia group** is spread over barely 1.4 Gm in semi-major axes, 1.6° in inclination (27.5 ± 0.8°), and eccentricities between 0.11 and 0.25. It has been suggested that the group could be a remnant of the break-up of an asteroid from the asteroid belt.
  - Carpo is another prograde moon and is not part of a known family. It has the highest inclination of all of the prograde moons.
  - S/2016 J 2, reported 2018, is the outermost prograde moon and is not part of a known family.
- Retrograde satellites:
  - The **Carme group** is spread over only 1.2 Gm in semi-major axis, 1.6° in inclination (165.7 ± 0.8°), and eccentricities between 0.23 and 0.27. It is very homogeneous in color (light red) and is believed to have originated from a D-type asteroid progenitor, possibly a Jupiter Trojan.
  - The **Ananke group** has a relatively wider spread than the previous groups, over 2.4 Gm in semi-major axis, 8.1° in inclination (between 145.7° and 154.8°), and eccentricities between 0.02 and 0.28. Most of the members appear gray, and are believed to have formed from the breakup of a captured asteroid.

- The **Pasiphae group** is quite dispersed, with a spread over 1.3 Gm, inclinations between 144.5° and 158.3°, and eccentricities between 0.25 and 0.43. The colors also vary significantly, from red to grey, which might be the result of multiple collisions. Sinope, sometimes included in the Pasiphae group, is red and, given the difference in inclination, it could have been captured independently; Pasiphae and Sinope are also trapped in secular resonances with Jupiter.

List

Key		
♠	†	‡
Galilean moons	Prograde irregular moons	Retrograde moons

The moons of Jupiter are listed below by orbital period. Moons massive enough for their surfaces to have collapsed into a spheroid are highlighted in bold. These are the four Galilean moons, which are comparable in size to the Moon. The other moons are much smaller, with the least massive Galilean moon being more than 7000 times more massive than the most massive of the other moons. The irregular captured moons are shaded light gray when prograde and dark gray when retrograde. All orbits are based on the estimated orbit on the Julian date 2457000, or 3 September 2017. As several moons of Jupiter are currently lost, these orbital elements may be only rough approximations. As of 2018, five satellites are considered to be lost. These are S/2003 J 2, S/2003 J 4, S/2003 J 10, S/2003 J 12, and S/2011 J 1. A number of other moons have only been observed for a year or two, but have decent enough orbits to be easily measurable even in 2018.

Order 462	Label 463	Name	Pronunciation	Image	Abs. magn.	Diameter (km) 464	Mass (×10^16 kg)	Semi-major axis (km)	Orbital period (d) 465	Inclination (°)	Eccentr.	Discovery year	Discoverer	Group 466
1	XVI	Metis	/ˈmiːtɪs/		10.5	60 × 40 × 34	≈3.6	128852	+7h 10m 16s	2.226	0.0077	1979	Synnott (Voyager 1)	Inner
2	XV	Adrastea	/əˈdræstiə/		12.0	20 × 16 × 14	≈0.2	129 000	+7h 15m 21s	2.217	0.0063	1979	Jewitt (Voyager 2)	Inner
3	V	Amalthea	/əˈmælθiə/ 467		7.1	250 × 146 × 128 (167±4.0)	208	181366	+12h 01m 46s	2.565	0.0075	1892	Barnard	Inner
4	XIV	Thebe	/ˈθiːbi/		9.0	116 × 98 × 84	≈43	222452	+16h 16m 02s	2.909	0.0180	1979	Synnott (Voyager 1)	Inner
5	I	Io♠	/ˈaɪoʊ/		-1.7	3660.0 × 3637.4 × 3630.6	8931900	421700	+1.7691	0.050	0.0041	1610	Galilei	Galilean
6	II	Europa♠	/fjʊəˈroʊpə/ 468		-1.4	3121.6	4800000	671034	+3.5512	0.471	0.0094	1610	Galilei	Galilean
7	III	Ganymede♠	/ˈgænɪmiːd/ 469,470		-2.1	5262.4	14819000	1070412	+7.1546	0.204	0.0011	1610	Galilei	Galilean

8	IV	Callisto♣	/kə'lɪstoʊ/	-1.2		4820.6	10759000	1882709	+16.689	0.205	0.0074	1610	Galilei	Galilean
9	XVIII	Themisto†	/θɪ'mɪstoʊ/	13.5		8	0.069	7393216	+129.87	45.762	0.2115	1975/-2000	Kowal & Roemer/ Sheppard et al.	*Themisto*
10	XIII	Leda†	/'liːdə/	12.8		16	0.6	11187781	+240.82	27.562	0.1673	1974	Kowal	Himalia
11	VI	Himalia†	/haɪ'meɪliə/	8.3		170	670	11451971	+250.23	30.486	0.1513	1904	Perrine	Himalia
12	—	S/2018 J 1†		15.9		2	0.0015	11453004	+250.40	30.606	0.0944	2018	Sheppard et al.	Himalia
13	—	S/2017 J 4†		16.2		2	0.0015	11494801	+251.77	28.155	0.1800	2017	Sheppard et al.	Himalia
14	X	Lysithea†	/laɪ'sɪθiə/	11.3		36	6.3	11740560	+259.89	27.006	0.1322	1938	Nicholson	Himalia
15	VII	Elara†	/'ɛlərə/	9.9		86	87	11778034	+259.64	29.691	0.1948	1905	Perrine	Himalia
16	LIII	Dia†	/'daɪə/	16.3		4	0.0090	12570424	+287.93	27.584	0.2058	2001	Sheppard et al.	Himalia

17	XLVI	Carpo†	/ˈkɑːrpoʊ/	16.2		3	0.0045	17144873	+458.62	56.001	0.2735	2003	Shep-pard et al.	Carpo
18	—(lost)	S/2003 J 12‡		17.0		1	0.00015	17739539 (28717431±1136944)	-482.69 (-944.29)	142.680 (152.5±1.3)	0.4449 (0.115±0.011)	2003	Shep-pard et al.	Ananke (unconfirmed)
19	—	S/2016 J 2†		16.9		1		18928095	+532.00	34.014	0.2219	2016	Shep-pard et al.	S/2016 J 2
20	XXXIV	Euporie‡	/juːˈpɔːriː/	16.4		2	0.0015	19088434	-538.78	144.694	0.0960	2002	Shep-pard et al.	Ananke
21	LX	S/2003 J 3‡		16.9		2	0.0015	19621780	-561.52	146.363	0.2507	2003	Shep-pard et al.	Ananke
22	—(lost)	S/2011 J 1‡		16.7		1		20155290 (24294975±113395)	-582.22 (-773.69)	162.8 (164.324±0.030)	0.2963 (0.1361±0.0011)	2011	Shep-pard et al.	Carme
23	LV	S/2003 J 18‡		16.5		2	0.0015	20219648	-587.38	146.376	0.1048	2003	Glad-man et al.	Ananke
24	LII	S/2010 J 2‡		17.5		1		20307150	-588.36	150.4	0.307	2010	Veillet	Ananke
25	XLII	Thelxinoe‡	/θɛlkˈsɪnoʊiː/	16.4		2	0.0015	20453753	-597.61	151.292	0.2684	2003	Shep-pard et al.	Ananke

No.	Designation	Name	Pronunciation	Mag	n		a (km)	Period	Inclination	Eccentricity	Year	Discoverer	Group
26	XXXIII	Euanthe‡	/juˈænθi/	16.5	3	0.0045	20464854	−598.09	143.409	0.2000	2002	Sheppard et al.	Ananke
27	XLV	Helike‡	/ˈhɛliki/	16.1	4	0.0090	20540266	−601.40	154.586	0.1374	2003	Sheppard et al.	Ananke
28	XXXV	Orthosie‡	/ɔːrˈθosiː/	16.7	2	0.0015	20567971	−602.62	142.366	0.2433	2002	Sheppard et al.	Ananke
29	—	S/2017 J 7‡		16.6	2	0.0015	20571458	−602.77	143.438	0.2147	2017	Sheppard et al.	Ananke
30	LIV	S/2016 J 1‡		17.0	3	0.0015	20595483	−603.83	139.839	0.1377	2016	Sheppard et al.	Ananke
31	—	S/2017 J 3‡		16.5	2	0.0015	20639315	−605.76	147.915	0.1477	2017	Sheppard et al.	Ananke
32	XXIV	Iocaste‡	/anooˈkæsti/	15.5	5	0.019	20722566	−609.43	147.248	0.2874	2001	Sheppard et al.	Ananke
33	—	S/2003 J 16‡		16.4	2	0.0015	20743779	−610.36	150.769	0.3184	2003	Gladman et al.	Ananke
34	XXVII	Praxidike‡	/-prekˈsɪdɪki/	14.9	7	0.043	20823948	−613.90	144.205	0.1840	2001	Sheppard et al.	Ananke

35	XXII	Harpalyke‡	/- hɑːrˈpælɪki/		15.9	4	0.012	21063814	−624.54	147.223	0.2440	2001	Sheppard et al.	Ananke
36	XL	Mneme‡	/ˈniːmi/		16.4	2	0.0015	21129786	−627.48	149.732	0.3169	2003	Gladman et al.	Ananke
37	XXX	Hermippe‡	/hərˈmɪpi/		15.6	4	0.0090	21182086	−629.81	151.242	0.2290	2002	Sheppard et al.	Ananke
38	XXIX	Thyone‡	/θaɪˈoʊni/		15.9	4	0.0090	21405570	−639.80	147.276	0.2525	2002	Sheppard et al.	Ananke
39	—	S/2017 J 9‡			16.1	2	0.0015	21429955	−640.90	152.661	0.2288	2017	Sheppard et al.	Ananke
40	XII	Ananke‡	/əˈnæŋki/		12.0	28	3.0	21454952	−640.38	151.564	0.3445	1951	Nicholson	Ananke
41	L	Herse‡	/ˈhɜːrsi/		16.6	2	0.0015	22134306	−672.75	162.490	0.2379	2003	Gladman et al.	Carme
42	XXXI	Aitne‡	/ˈetni/		16.0	3	0.0045	22285161	−679.64	165.562	0.3927	2002	Sheppard et al.	Carme
43	—	S/2017 J 6‡			16.4	2	0.0015	22394682	−684.66	155.185	0.5569	2017	Sheppard et al.	Pasiphae (fringe member)

#	Designation	Name	Pronunciation	Mag								Year	Discoverer	Group
44	XXXVII	Kale‡	/ˈkeɪli/	16.4		2	0.0015	22409207	−685.32	165.378	0.2011	2002	Sheppard et al.	Carme
45	XX	Taygete‡	/teɪˈɪdʒti/	15.6		5	0.016	22438648	−686.67	164.890	0.3678	2001	Sheppard et al.	Carme
46	—	S/2003 J 19‡		16.8		2	0.0015	22696750	−698.55	166.657	0.2572	2003	Gladman et al.	Carme
47	XXI	Chaldene‡	/kælˈdiːni/	16.0		4	0.0075	22713444	−699.33	167.070	0.2916	2001	Sheppard et al.	Carme
48	LVIII	S/2003 J 15‡		16.7		2	0.0015	22720999	−699.68	141.812	0.0932	2003	Sheppard et al.	Pasiphae
49	—(lost)	S/2003 J 10‡		16.8		2	0.0015	22730813 (22462575±670198)	−700.13 (−687.83)	163.813 (162.38±0.90)	0.3438 (0.095±0.014)	2003	Sheppard et al.	Carme
50	—	S/2003 J 23‡		16.8		2	0.0015	22739654	−700.54	148.849	0.3930	2004	Sheppard et al.	Pasiphae
51	XXV	Erinome‡	/ɪˈrɪnoʊmi/	16.1		3	0.0045	22986266	−711.96	163.737	0.2552	2001	Sheppard et al.	Carme
52	XLI	Aoede‡	/eɪˈiːdi/	15.6		4	0.0090	23044175	−714.66	160.482	0.4311	2003	Sheppard et al.	Pasiphae

53	XLIV	Kallichore‡	/ˈkalikoːri/	16.3	2	0.0015	23111823	−717.81	164.605	0.2041	2003	Sheppard et al.	Carme
54	—	S/2017 J 5‡		16.5	2	0.0015	23169389	−720.49	164.331	0.2842	2017	Sheppard et al.	Carme
55	—	S/2017 J 8‡		17.0	1	0.0015	23174446	−720.73	164.782	0.3118	2017	Sheppard et al.	Carme
56	XXIII	Kalyke‡	/ˈkæliki/	15.5	5	0.019	23180773	−721.02	165.505	0.2139	2001	Sheppard et al.	Carme
57	XI	Carme‡	/ˈkɑːrmi/	11.0	46	13	23197992	−702.28	165.047	0.2342	1938	Nicholson	Carme
58	XVII	Callirrhoe‡	/kəˈlɪroʊ/	14.1	9	0.087	23214986	−727.11	139.849	0.2582	2000	Spahr, Scotti	Pasiphae
59	XXXII	Eurydome‡	/jʊəˈrɪdəmi/	16.3	3	0.0045	23230858	−723.36	149.324	0.3769	2002	Sheppard et al.	Pasiphae
60	—	S/2017 J 2‡		16.9	2	0.0015	23240957	−723.83	166.398	0.2360	2017	Sheppard et al.	Carme
61	XXXVIII	Pasithee‡	/ˈpasɪθiː/	16.8	2	0.0015	23307318	−726.93	165.759	0.3288	2002	Sheppard et al.	Carme

#	Name	Pronunciation		Mag			Semi-major axis	Period	Inclination	Eccentricity	Year	Discoverer	Group
62	S/2010 J1‡			16.5	2		23314335	−724.34	163.2	0.320	2010	Jacobson et al.	Carme
63	Kore‡	/ˈkɔːri/		16.6	2	0.0015	23345093	−723.72	137.371	0.1951	2003	Sheppard et al.	Pasiphae
64	Cyllene‡	/sɪˈliːni/		16.3	2	0.0015	23396269	−731.10	140.148	0.4115	2003	Sheppard et al.	Pasiphae
65	S/2011 J2‡			16.9	1		23400981	−731.32	148.77	0.3321	2011	Sheppard et al.	Pasiphae
66	Eukelade‡	/juːˈkɛlədi/		16.0	4	0.0090	23483694	−735.20	163.996	0.2828	2003	Sheppard et al.	Carme
67	S/2017 J1‡			16.8	2	0.0015	23483978	−734.15	149.197	0.3969	2017	Sheppard et al.	Pasiphae
68	— (lost)	S/2003 J4‡		16.7	2	0.0015	23570790 (22766748±1780215)	−739.29 (−701.85)	147.175 (143.2±1.3)	0.3003 (0.1111±0.0077)	2003	Sheppard et al.	Pasiphae
69	Pasiphae‡	/pəˈsɪfaɪi/		10.4	60	30	23609042	−739.80	141.803	0.3743	1908	Melotte	Pasiphae
70	Hegemone‡	/hɪˈdʒɛməni/		16.0	3	0.0045	23702511	−745.50	152.506	0.4077	2003	Sheppard et al.	Pasiphae
71	Arche‡	/ˈɑːrki/		16.3	3	0.0045	23717051	−746.19	164.587	0.1492	2002	Sheppard et al.	Carme

72	XXVI	Isonoe‡	/aiˈsɒnooi/	16.0		4	0.0075	23800647	−750.13	165.127	0.1775	2001	Sheppard et al.	Carme
73	—	S/2003 J 9‡		17.0		1	0.00015	23857808	−752.84	164.980	0.2761	2003	Sheppard et al.	Carme
74	LVII	S/2003 J 5‡		15.9		4	0.0090	23973926	−758.34	165.549	0.3070	2003	Sheppard et al.	Carme
75	IX	Sinope‡	/siˈnoopi/	11.4		38	7.5	24057865	−739.33	153.778	0.2750	1914	Nicholson	Pasiphae
76	XXXVI	Sponde‡	/ˈspɒndi/	16.7		2	0.0015	24252627	−771.60	154.372	0.4431	2002	Sheppard et al.	Pasiphae
77	XXVIII	Autonoe‡	/ɔːˈtɒnooi/	15.6		4	0.0090	24264445	−772.17	151.058	0.3690	2002	Sheppard et al.	Pasiphae
78	XIX	Megaclite‡	/ˈmega ˈklati/	15.0		5	0.021	24687239	−792.44	150.398	0.3077	2001	Sheppard et al.	Pasiphae
79	— (lost)	S/2003 J 2‡		16.6		2	0.0015	28570410 (27734694±10756087)	−981.55 (−943.69)	153.521 (151.3±2.5)	0.4074 (0.1197±0.0024)	2003	Sheppard et al.	Pasiphae (unconfirmed)

Exploration

The first spacecrafts to visit Jupiter were *Pioneer 10* in 1973, and *Pioneer 11* a year later, taking low-resolution images of the four Galilean moons.[471] The *Voyager 1* and *Voyager 2* probes visited Jupiter in 1979, discovering the volcanic activity on Io and the presence of water ice on the surface of Europa. The *Cassini* probe to Saturn flew by Jupiter in 2000 and collected data on interactions of the Galilean moons with Jupiter's extended atmosphere. The *New Horizons* spacecraft flew by Jupiter in 2007 and made improved measurements of its satellites' orbital parameters.

The *Galileo* spacecraft was the first to enter orbit around Jupiter, arriving in 1995 and studying it until 2003. During this period, *Galileo* gathered a large amount of information about the Jovian system, making close approaches to all of the Galilean moons and finding evidence for thin atmospheres on three of them, as well as the possibility of liquid water beneath the surfaces of Europa, Ganymede, and Callisto. It also discovered a magnetic field around Ganymede.

In 2016, the *Juno* spacecraft imaged the Galilean moons from above their orbital plane as it approached Jupiter orbit insertion, creating a time-lapse movie of their motion.[472]

External links

> Wikimedia Commons has media related to *Moons of Jupiter*.

- Jupiter's Known Satellites[473]
- The Jupiter Satellite and Moon Page[474]
- Simulation showing the position of Jupiter's moons[475]
- Animated tour of Jupiter's moons[476], University of South Wales
- Jupiter:Moons[477] by NASA's Solar System Exploration[478]
- David Perlman (15 May 2003). "43 more moons orbiting Jupiter"[479]. *San Francisco Chronicle*.
- Archive of Jupiter System Articles[480] in Planetary Science Research Discoveries[481]
- An animation of the Jovian system of moons[482]
- Cain, Fraser: Astronomy Cast Ep 57: Jupiter's Moons[483]

<indicator name="featured-star"> ⭐ </indicator>

Galilean moons

Galilean moons

The **Galilean moons** are the four largest moons of Jupiter—Io, Europa, Ganymede, and Callisto. They were first seen by Galileo Galilei in January 1610, and recognized by him as satellites of Jupiter in March 1610.[484] They were the first objects found to orbit another planet. Their names derive from the lovers of Zeus. They are among the largest objects in the Solar System with the exception of the Sun and the eight planets, with a radius larger than any of the dwarf planets. Ganymede is the largest moon in the Solar System, and is even bigger than the planet Mercury, though only around half as massive. The three inner moons—Io, Europa, and Ganymede—are in a 4:2:1 orbital resonance with each other. Because of their much smaller size, and therefore weaker self-gravitation, all of Jupiter's remaining moons have irregular forms rather than a spherical shape.

The Galilean moons were discovered in either 1609 or 1610 when Galileo made improvements to his telescope, which enabled him to observe celestial bodies more distinctly than ever. Galileo's discovery showed the importance of the telescope as a tool for astronomers by proving that there were objects in space that cannot be seen by the naked eye. More importantly, the incontrovertible discovery of celestial bodies orbiting something other than Earth dealt a serious blow to the then-accepted Ptolemaic world system, or the geocentric theory in which everything orbits around Earth.

Galileo initially named his discovery the **Cosmica Sidera** ("Cosimo's stars"), but the names that eventually prevailed were chosen by Simon Marius. Marius discovered the moons independently at the same time as Galileo, and gave them their present names, which were suggested by Johannes Kepler, in his *Mundus Jovialis*, published in 1614.

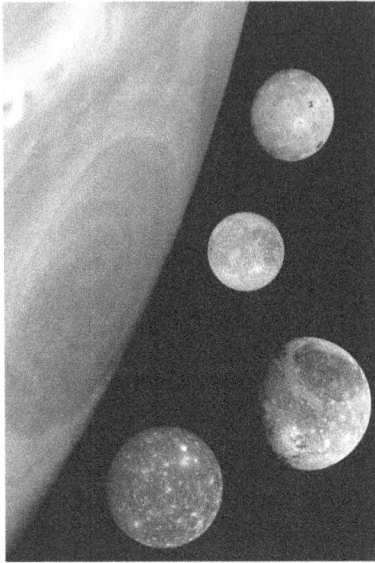

Figure 76: *Montage of Jupiter's four Galilean moons, in a composite image depicting part of Jupiter and their relative sizes (positions are illustrative, not actual). From top to bottom: Io, Europa, Ganymede, Callisto.*

Figure 77: *Two Hubble Space Telescope views of a rare triple transit of Jupiter by Europa, Callisto and Io (24 January 2015).*

Figure 78: *Galileo Galilei, the discoverer of the four moons*

History

Discovery

As a result of improvements Galileo Galilei made to the telescope, with a magnifying capability of $20\times$, he was able to see celestial bodies more distinctly than was ever possible before. This allowed Galilei to discover in either December 1609 or January 1610 what came to be known as the Galilean moons.[485]

On January 7, 1610, Galileo wrote a letter containing the first mention of Jupiter's moons. At the time, he saw only three of them, and he believed them to be fixed stars near Jupiter. He continued to observe these celestial orbs from January 8 to March 2, 1610. In these observations, he discovered a fourth body, and also observed that the four were not fixed stars, but rather were orbiting Jupiter.

Galileo's discovery proved the importance of the telescope as a tool for astronomers by showing that there were objects in space to be discovered that until then had remained unseen by the naked eye. More importantly, the discovery of celestial bodies orbiting something other than Earth dealt a blow to the then-accepted Ptolemaic world system, which held that Earth was at the center of the universe and all other celestial bodies revolved around it. Galileo's

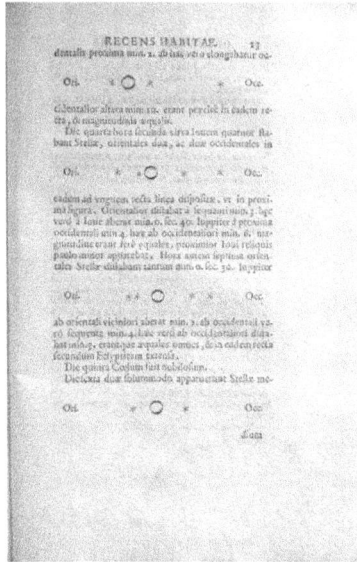

Figure 79: *The Medician stars in the Sidereus Nuncius (the 'starry messenger'), 1610. The moons are drawn in changing positions.*

Sidereus Nuncius (*Starry Messenger*), which announced celestial observations through his telescope, does not explicitly mention Copernican heliocentrism, a theory that placed the Sun at the center of the universe. Nevertheless, Galileo accepted the Copernican theory.

A Chinese historian of astronomy, Xi Zezong, has claimed that a "small reddish star" observed near Jupiter in 362 BCE by Chinese astronomer Gan De may have been Ganymede, predating Galileo's discovery by around two millennia.[486]

Dedication to the Medicis

In 1605, Galileo had been employed as a mathematics tutor for Cosimo de' Medici. In 1609, Cosimo became Grand Duke Cosimo II of Tuscany. Galileo, seeking patronage from his now-wealthy former student and his powerful family, used the discovery of Jupiter's moons to gain it. On February 13, 1610, Galileo wrote to the Grand Duke's secretary:

"God graced me with being able, through such a singular sign, to reveal to my Lord my devotion and the desire I have that his glorious name live as equal among the stars, and since it is up to me, the first discoverer, to name these new planets, I wish, in imitation of the great sages who placed

the most excellent heroes of that age among the stars, to inscribe these with the name of the Most Serene Grand Duke. "

Galileo asked whether he should name the moons the "Cosmian Stars", after Cosimo alone, or the "Medician Stars", which would honor all four brothers in the Medici clan. The secretary replied that the latter name would be best.

On March 12, 1610, Galileo wrote his dedicatory letter to the Duke of Tuscany, and the next day sent a copy to the Grand Duke, hoping to obtain the Grand Duke's support as quickly as possible. On March 19, he sent the telescope he had used to first view Jupiter's moons to the Grand Duke, along with an official copy of *Sidereus Nuncius (The Starry Messenger)* that, following the secretary's advice, named the four moons the Medician Stars. In his dedicatory introduction, Galileo wrote:

Scarcely have the immortal graces of your soul begun to shine forth on earth than bright stars offer themselves in the heavens which, like tongues, will speak of and celebrate your most excellent virtues for all time. Behold, therefore, four stars reserved for your illustrious name ... which ... make their journeys and orbits with a marvelous speed around the star of Jupiter ... like children of the same family ... Indeed, it appears the Maker of the Stars himself, by clear arguments, admonished me to call these new planets by the illustrious name of Your Highness before all others.

Name

Galileo initially called his discovery the **Cosmica Sidera** ("Cosimo's stars"), in honour of Cosimo II de' Medici (1590–1621).[487] At Cosimo's suggestion, Galileo changed the name to **Medicea Sidera** ("the **Medician stars**"), honouring all four Medici brothers (Cosimo, Francesco, Carlo, and Lorenzo). The discovery was announced in the *Sidereus Nuncius* ("Starry Messenger"), published in Venice in March 1610, less than two months after the first observations.

Other names put forward include:

- I. *Principharus* (for the "prince" of Tuscany), II. *Victripharus* (after Vittoria della Rovere), III. *Cosmipharus* (after Cosimo de' Medici) and IV. *Fernipharus* (after Duke Ferdinando de' Medici) – by Giovanni Battista Hodierna, a disciple of Galileo and author of the first ephemerides (*Medicaeorum Ephemerides*, 1656);
- *Circulatores Jovis*, or *Jovis Comites* – by Johannes Hevelius;
- *Gardes*, or *Satellites* (from the Latin *satelles, satellitis*, meaning "escorts") – by Jacques Ozanam.

Figure 80: *A Jovilabe: an apparatus from the mid-18th century for demonstrating the orbits of Jupiter's satellites*

The names that eventually prevailed were chosen by Simon Marius, who discovered the moons independently at the same time as Galileo: he named them at the suggestion of Johannes Kepler after lovers of the god Zeus (the Greek equivalent of Jupiter): *Io, Europa, Ganymede* and *Callisto*, in his *Mundus Jovialis*, published in 1614.

Galileo steadfastly refused to use Marius' names and invented as a result the numbering scheme that is still used nowadays, in parallel with proper moon names. The numbers run from Jupiter outward, thus I, II, III and IV for Io, Europa, Ganymede, and Callisto respectively. Galileo used this system in his notebooks but never actually published it. The numbered names (Jupiter *x*) were used until the mid-20th century when other inner moons were discovered, and Marius' names became widely used.

Determination of longitude

Galileo was able to develop a method of determining longitude based on the timing of the orbits of the Galilean moons.[488] The times of the eclipses of the moons could be precisely calculated in advance, and compared with local observations on land or on ship to determine the local time and hence longitude. The main problem with the technique was that it was difficult to observe the

Galilean moons through a telescope on a moving ship; a problem that Galileo tried to solve with the invention of the celatone. The method was used by Cassini and Picard to re-map France.

Members

Some models predict that there may have been several generations of Galilean satellites in Jupiter's early history. Each generation of moons to have formed would have spiraled into Jupiter and been destroyed, due to tidal interactions with Jupiter's proto-satellite disk, with new moons forming from the remaining debris. By the time the present generation formed, the gas in the proto-satellite disk had thinned out to the point that it no longer greatly interfered with the moons' orbits. Other models suggest that Galilean satellites formed in a proto-satellite disk, in which formation timescales were comparable to or shorter than orbital migration timescales. Io is anhydrous and likely has an interior of rock and metal. Europa is thought to contain 8% ice and water by mass with the remainder rock. These moons are, in increasing order of distance from Jupiter:

Name	Image	Model of interior	Diameter (km)	Mass (kg)	Density (g/cm³)	Semi-major axis (km)[489]	Orbital period(d)[490] (relative)	Inclination (°)[491]	Eccentricity
Io *Jupiter I*			3660.0 × 3637.4 × 3630.6	8.93×10^{22}	3.528	421800	1.769 (1)	0.050	0.0041
Europa *Jupiter II*			3121.6	4.8×10^{22}	3.014	671100	3.551 (2)	0.471	0.0094
Ganymede *Jupiter III*			5268.2	1.48×10^{23}	1.942	1070400	7.155 (4)	0.204	0.0011
Callisto *Jupiter IV*			4820.6	1.08×10^{23}	1.834	1882700	16.69 (9.4)	0.205	0.0074

Io

Io (Jupiter I) is the innermost of the four Galilean moons of Jupiter and, with a diameter of 3642 kilometers, the fourth-largest moon in the Solar System. It was named after Io, a priestess of Hera who became one of the lovers of Zeus. Nevertheless, it was simply referred to as "Jupiter I", or "The first satellite of Jupiter", until the mid-20th century.

With over 400 active volcanos, Io is the most geologically active object in the Solar System. Its surface is dotted with more than 100 mountains, some of which are taller than Earth's Mount Everest. Unlike most satellites in the outer Solar System (which have a thick coating of ice), Io is primarily composed of silicate rock surrounding a molten iron or iron sulfide core.

Although not proven, recent data from the Galileo orbiter indicate that Io might have its own magnetic field. Io has an extremely thin atmosphere made up mostly of sulfur dioxide (SO_2). If a surface data or collection vessel were to land on Io in the future, it would have to be extremely tough (similar to the tank-like bodies of the Soviet Venera landers) to survive the radiation and magnetic fields that originate from Jupiter.

Europa

Europa (Jupiter II), the second of the four Galilean moons, is the second closest to Jupiter and the smallest at 3121.6 kilometers in diameter, which is slightly smaller than the Moon. The name comes from a mythical Phoenician noblewoman, Europa, who was courted by Zeus and became the queen of Crete, though the name did not become widely used until the mid-20th century.

It has a smooth and bright surface, with a layer of water surrounding the mantle of the planet, thought to be 100 kilometers thick.[492] The smooth surface includes a layer of ice, while the bottom of the ice is theorized to be liquid water. The apparent youth and smoothness of the surface have led to the hypothesis that a water ocean exists beneath it, which could conceivably serve as an abode for extraterrestrial life. Heat energy from tidal flexing ensures that the ocean remains liquid and drives geological activity. Life may exist in Europa's under-ice ocean. So far, there is no evidence that life exists on Europa, but the likely presence of liquid water has spurred calls to send a probe there.

The prominent markings that criss-cross the moon seem to be mainly albedo features, which emphasize low topography. There are few craters on Europa because its surface is tectonically active and young. Some theories suggest that Jupiter's gravity is causing these markings, as one side of Europa is constantly facing Jupiter. Also, volcanic water eruptions splitting the surface of Europa, and even geysers have been considered as a cause. The color of the markings,

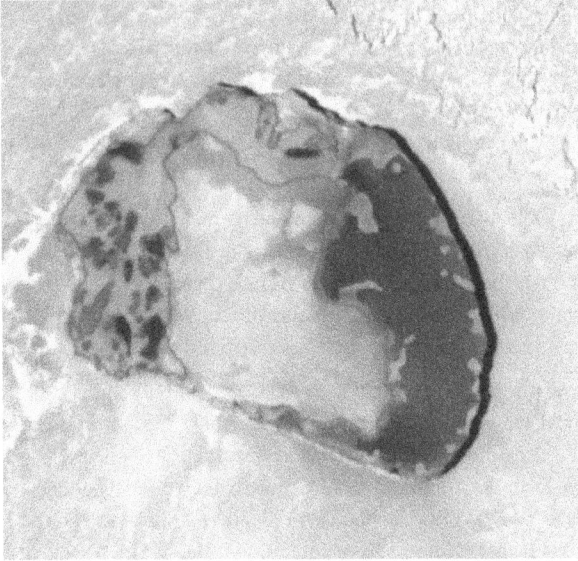

Figure 81:
Tupan Patera on Io

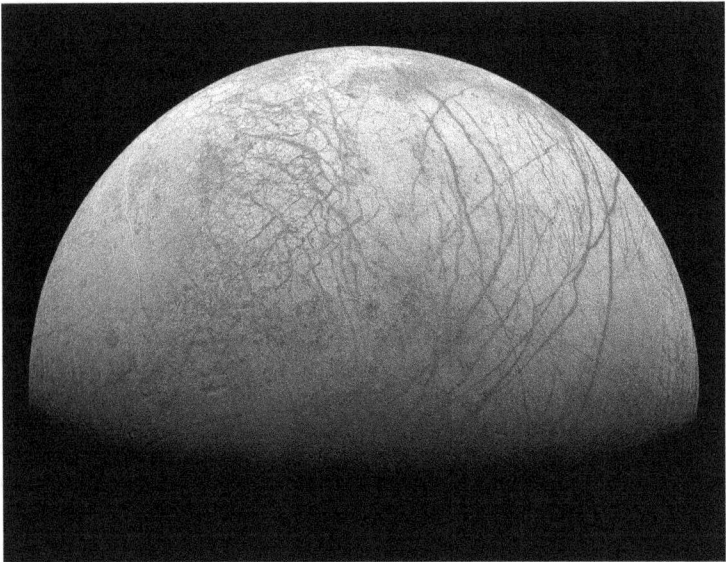

Figure 82: *Europa*

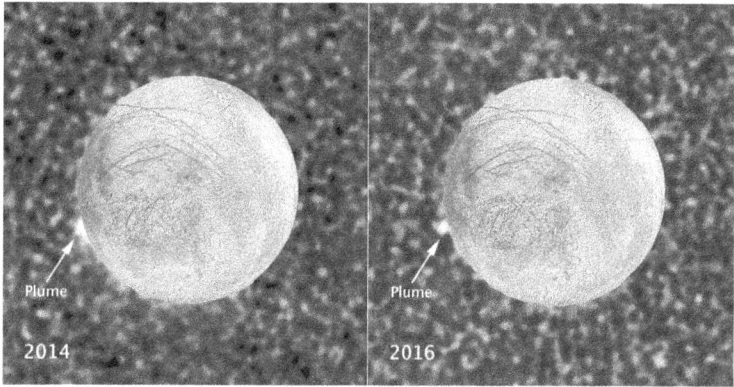

Figure 83: *Recurring plume erupting from Europa.*

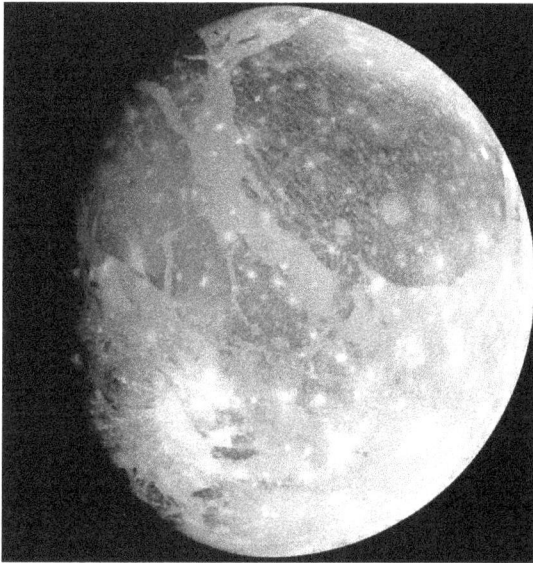

Figure 84: *Ganymede*

reddish-brown, is theorized to be caused by sulfur, but scientists cannot confirm that, because no data collection devices have been sent to Europa. Europa is primarily made of silicate rock and likely has an iron core. It has a tenuous atmosphere composed primarily of oxygen.

Figure 85: *Callisto's Valhalla impact crater as seen by Voyager*

Ganymede

Ganymede (Jupiter III), the third Galilean moon is named after the mythological Ganymede, cupbearer of the Greek gods and Zeus's beloved. Ganymede is
the largest natural satellite in the Solar System at 5262.4 kilometers in diameter, which makes it larger than the planet Mercury – although only at about half
of its mass since Ganymede is an icy world. It is the only satellite in the Solar
System known to possess a magnetosphere, likely created through convection
within the liquid iron core.

Ganymede is composed primarily of silicate rock and water ice, and a salt-
water ocean is believed to exist nearly 200 km below Ganymede's surface,
sandwiched between layers of ice. The metallic core of Ganymede suggests
a greater heat at some time in its past than had previously been proposed.
The surface is a mix of two types of terrain—highly cratered dark regions and
younger, but still ancient, regions with a large array of grooves and ridges.
Ganymede has a high number of craters, but many are gone or barely visible due to its icy crust forming over them. The satellite has a thin oxygen
atmosphere that includes O, O_2, and possibly O_3 (ozone), and some atomic
hydrogen.

Callisto

Callisto (Jupiter IV) is the fourth and last Galilean moon, and is the second
largest of the four, and at 4820.6 kilometers in diameter, it is the third largest
moon in the Solar System, and barely smaller than Mercury, though only a

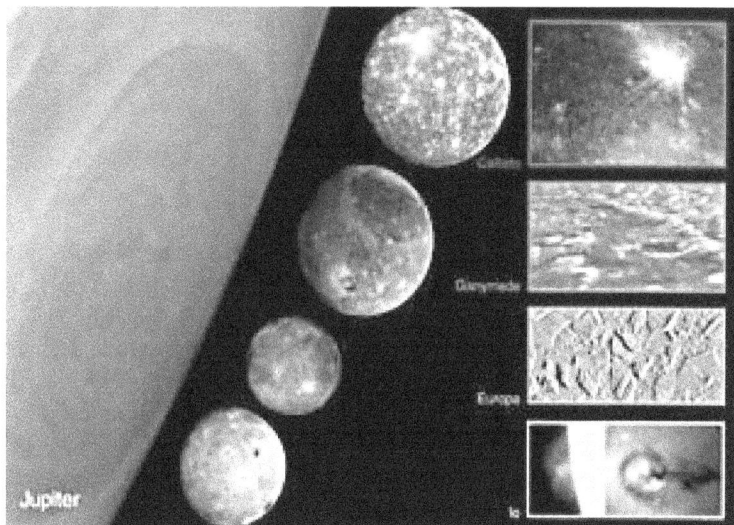

Figure 86: *Comparison of (a part of) Jupiter and its four largest natural satellites*

third of the latter's mass. It is named after the Greek mythological nymph Callisto, a lover of Zeus who was a daughter of the Arkadian King Lykaon and a hunting companion of the goddess Artemis. The moon does not form part of the orbital resonance that affects three inner Galilean satellites and thus does not experience appreciable tidal heating. Callisto is composed of approximately equal amounts of rock and ices, which makes it the least dense of the Galilean moons. It is one of the most heavily cratered satellites in the Solar System, and one major feature is a basin around 3000 km wide called Valhalla.

Callisto is surrounded by an extremely thin atmosphere composed of carbon dioxide and probably molecular oxygen. Investigation revealed that Callisto may possibly have a subsurface ocean of liquid water at depths less than 300 kilometres. The likely presence of an ocean within Callisto indicates that it can or could harbour life. However, this is less likely than on nearby Europa. Callisto has long been considered the most suitable place for a human base for future exploration of the Jupiter system since it is furthest from the intense radiation of Jupiter.

Comparative structure

Jovian Radiation

Moon	rem/day
Io	3600
Europa	540
Ganymede	8
Callisto	0.01

Fluctuations in the orbits of the moons indicate that their mean density decreases with distance from Jupiter. Callisto, the outermost and least dense of the four, has a density intermediate between ice and rock whereas Io, the innermost and densest moon, has a density intermediate between rock and iron. Callisto has an ancient, heavily cratered and unaltered ice surface and the way it rotates indicates that its density is equally distributed, suggesting that it has no rocky or metallic core but consists of a homogeneous mix of rock and ice. This may well have been the original structure of all the moons. The rotation of the three inner moons, in contrast, indicates differentiation of their interiors with denser matter at the core and lighter matter above. They also reveal significant alteration of the surface. Ganymede reveals past tectonic movement of the ice surface which required partial melting of subsurface layers. Europa reveals more dynamic and recent movement of this nature, suggesting a thinner ice crust. Finally, Io, the innermost moon, has a sulfur surface, active volcanism and no sign of ice. All this evidence suggests that the nearer a moon is to Jupiter the hotter its interior. The current model is that the moons experience tidal heating as a result of the gravitational field of Jupiter in inverse proportion to the square of their distance from the giant planet. In all but Callisto this will have melted the interior ice, allowing rock and iron to sink to the interior and water to cover the surface. In Ganymede a thick and solid ice crust then formed. In warmer Europa a thinner more easily broken crust formed. In Io the heating is so extreme that all the rock has melted and water has long ago boiled out into space.

Size

Latest flyby

<templatestyles src="Multiple_image/styles.css" />

Jupiter and Io

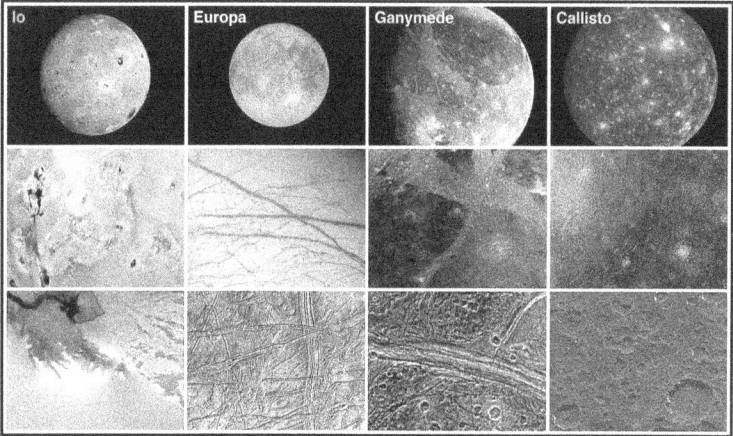

Figure 87:
Surface features of the four members at different levels of zoom in each row

Figure 88: *Galilean moons compared with other Solar System bodies, although pixel scale is not accurate at this resolution.*

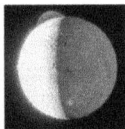

Io

Europa

Ganymede

Callisto

Jupiter and Galilean moons circa 2007, imaged
by *New Horizons* during flyby. (greyscale colour)

Origin and evolution

Jupiter's regular satellites are believed to have formed from a circumplanetary disk, a ring of accreting gas and solid debris analogous to a protoplanetary disk. They may be the remnants of a score of Galilean-mass satellites that formed early in Jupiter's history.

Simulations suggest that, while the disk had a relatively high mass at any given moment, over time a substantial fraction (several tenths of a percent) of the mass of Jupiter captured from the Solar nebula was processed through it. However, the disk mass of only 2% that of Jupiter is required to explain the existing satellites. Thus there may have been several generations of Galilean-mass satellites in Jupiter's early history. Each generation of moons would have spiraled into Jupiter, due to drag from the disk, with new moons then forming from the new debris captured from the Solar nebula. By the time the present (possibly fifth) generation formed, the disk had thinned out to the point that

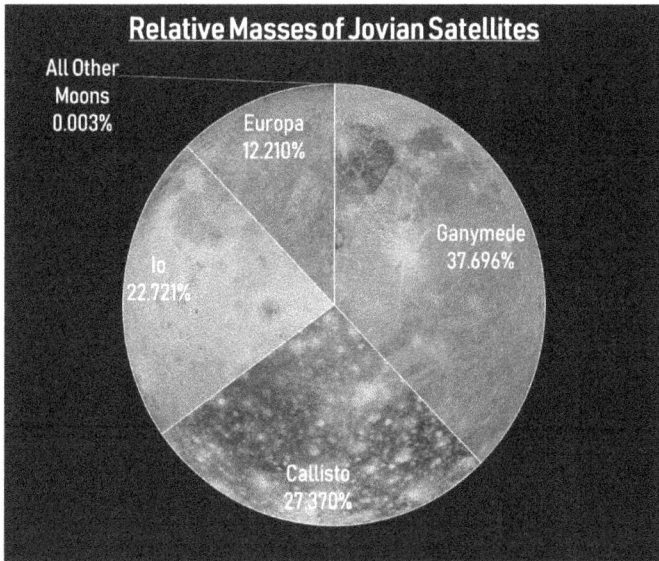

Figure 89: *The relative masses of the Jovian moons. Those smaller than Europa are not visible at this scale, and combined would only be visible at 100× magnification.*

it no longer greatly interfered with the moons' orbits. The current Galilean moons were still affected, falling into and being partially protected by an orbital resonance which still exists for Io, Europa, and Ganymede. Ganymede's larger mass means that it would have migrated inward at a faster rate than Europa or Io.

Visibility

All four Galilean moons are bright enough to be viewed from Earth without a telescope, if they appear farther away from Jupiter. (They are, however, easily visible with even low-powered binoculars.) They have apparent magnitudes between 4.6 and 5.6 when Jupiter is in opposition with the Sun, and are about one unit of magnitude dimmer when Jupiter is in conjunction. The main difficulty in observing the moons from Earth is their proximity to Jupiter, since they are obscured by its brightness.[493] The maximum angular separations of the moons are between 2 and 10 arcminutes from Jupiter,[494] which is close to the limit of human visual acuity. Ganymede and Callisto, at their maximum separation, are the likeliest targets for potential naked-eye observation.

Figure 90: *Jupiter and its four Galilean moons as seen with an amateur telescope*

Figure 91: *Jupiter with the Galilean moons – Io, Ganymede, Europa, and Callisto (near maximum elongation), respectively – and the full Moon as seen around conjunction on 10 April 2017*

GANYMEDE 4:1

EUROPA 2:1

IO 1:1

JUPITER

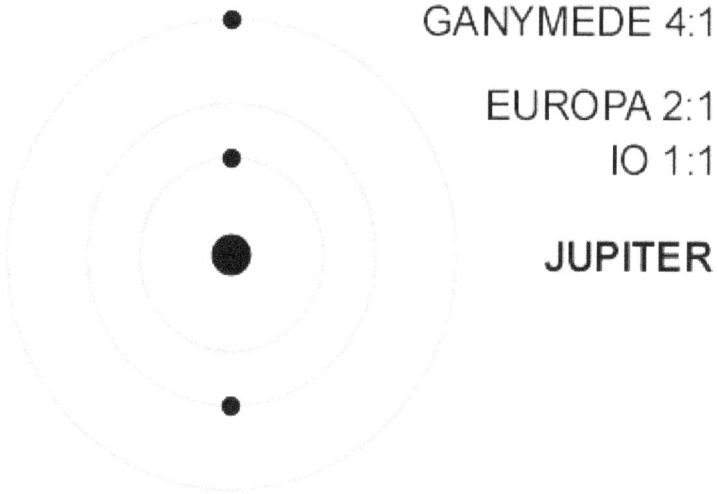

Figure 92: *The three inner Galilean moons revolve in a 1:2:4 resonance.*

Orbit animation

GIF animation of the resonance of Io, Europa, and Ganymede

External links

- Animation of Galileo's observation, march 1613[495]
- Sky & Telescope utility for identifying Galilean moons[496]

Wikimedia Commons has media related to *Moons of Jupiter*.

<indicator name="good-star"> ⊕ </indicator>

Planetary rings

Rings of Jupiter

The planet Jupiter has a system of rings known as the **rings of Jupiter** or the **Jovian ring system**. It was the third ring system to be discovered in the Solar System, after those of Saturn and Uranus. It was first observed in 1979 by the *Voyager 1* space probe and thoroughly investigated in the 1990s by the *Galileo* orbiter. It has also been observed by the Hubble Space Telescope and from Earth for several years. Ground-based observations of the rings require the largest available telescopes.

The Jovian ring system is faint and consists mainly of dust. It has four main components: a thick inner torus of particles known as the "halo ring"; a relatively bright, exceptionally thin "main ring"; and two wide, thick and faint outer "gossamer rings", named for the moons of whose material they are composed: Amalthea and Thebe.

The main and halo rings consist of dust ejected from the moons Metis, Adrastea, and other unobserved parent bodies as the result of high-velocity impacts. High-resolution images obtained in February and March 2007 by the *New Horizons* spacecraft revealed a rich fine structure in the main ring.

In visible and near-infrared light, the rings have a reddish color, except the halo ring, which is neutral or blue in color. The size of the dust in the rings varies, but the cross-sectional area is greatest for nonspherical particles of radius about 15 μm in all rings except the halo. The halo ring is probably dominated by submicrometre dust. The total mass of the ring system (including unresolved parent bodies) is poorly known, but is probably in the range of 10^{11} to 10^{16} kg. The age of the ring system is not known, but it may have existed since the formation of Jupiter.

A ring could possibly exist in Himalia's orbit. One possible explanation is that a small moon had crashed into Himalia and the force of the impact caused material to blast off Himalia.[497]

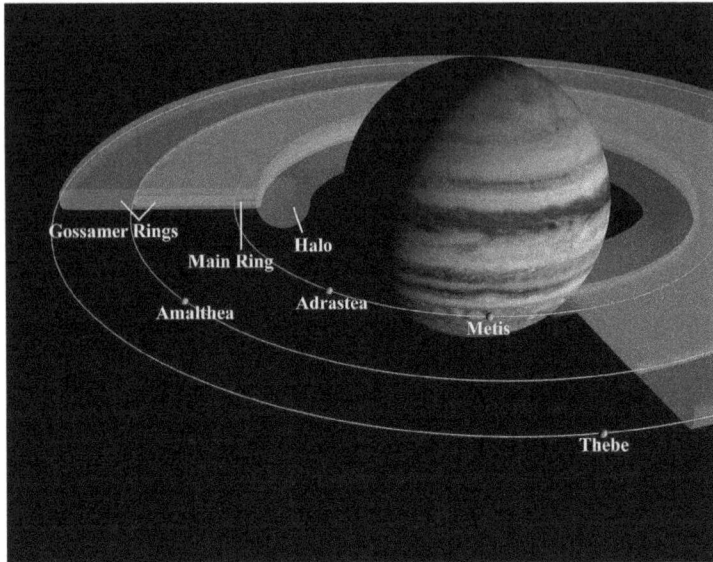

Figure 93: *A schema of Jupiter's ring system showing the four main components. For simplicity, Metis and Adrastea are depicted as sharing their orbit.*

Discovery and structure

Jupiter's ring system was the third to be discovered in the Solar System, after those of Saturn and Uranus. It was first observed in 1979 by the *Voyager 1* space probe. It is composed of four main components: a thick inner torus of particles known as the "halo ring"; a relatively bright, exceptionally thin "main ring"; and two wide, thick and faint outer "gossamer rings", named after the moons of whose material they are composed: Amalthea and Thebe. The principal attributes of the known Jovian Rings are listed in the table.

Name	Radius (km)	Width (km)	Thick-ness (km)	Optical depth[498] (in τ)	Dust frac-tion	Mass, kg	Notes
Halo ring	92000–122500	30500	12500	$\sim 1 \times 10^{-6}$	100%	—	
Main ring	122500–129000	6500	30–300	5.9×10^{-6}	$\sim 25\%$	10^7–10^9 (dust) 10^{11}–10^{16} (large parti-cles)	Bounded by Adrastea

Amalthea gossamer ring	129000–182000	53000	2000	~1 × 10^{-7}	100%	10^7– 10^9	Connected with Amalthea
Thebe gossamer ring	129000–226000	97000	8400	~3 × 10^{-8}	100%	10^7– 10^9	Connected with Thebe. There is an extension beyond the orbit of Thebe.

Main ring

Appearance and structure

The narrow and relatively thin main ring is the brightest part of Jupiter's ring system. Its outer edge is located at a radius of about 129000 km (1.806 R_J; R_J = equatorial radius of Jupiter or 71398 km) and coincides with the orbit of Jupiter's smallest inner satellite, Adrastea. Its inner edge is not marked by any satellite and is located at about 122500 km (1.72 R_J).

Thus the width of the main ring is around 6500 km. The appearance of the main ring depends on the viewing geometry. In forward-scattered light[499] the brightness of the main ring begins to decrease steeply at 128600 km (just inward of the Adrastean orbit) and reaches the background level at 129300 km—just outward of the Adrastean orbit. Therefore, Adrastea at 129000 km clearly shepherds the ring. The brightness continues to increase in the direction of Jupiter and has a maximum near the ring's center at 126000 km, although there is a pronounced gap (notch) near the Metidian orbit at 128000 km. The inner boundary of the main ring, in contrast, appears to fade off slowly from 124000 to 120000 km, merging into the halo ring. In forward-scattered light all Jovian rings are especially bright.

In back-scattered light[500] the situation is different. The outer boundary of the main ring, located at 129100 km, or slightly beyond the orbit of Adrastea, is very steep. The orbit of the moon is marked by a gap in the ring so there is a thin ringlet just outside its orbit. There is another ringlet just inside Adrastean orbit followed by a gap of unknown origin located at about 128500 km. The third ringlet is found inward of the central gap, outside the orbit of Metis. The ring's brightness drops sharply just outward of the Metidian orbit, forming the Metis notch. Inward of the orbit of Metis, the brightness of the ring rises much less than in forward-scattered light. So in the back-scattered geometry the main ring appears to consist of two different parts: a narrow outer part extending from 128000 to 129000 km, which itself includes three narrow ringlets separated by notches, and a fainter inner part from 122500 to

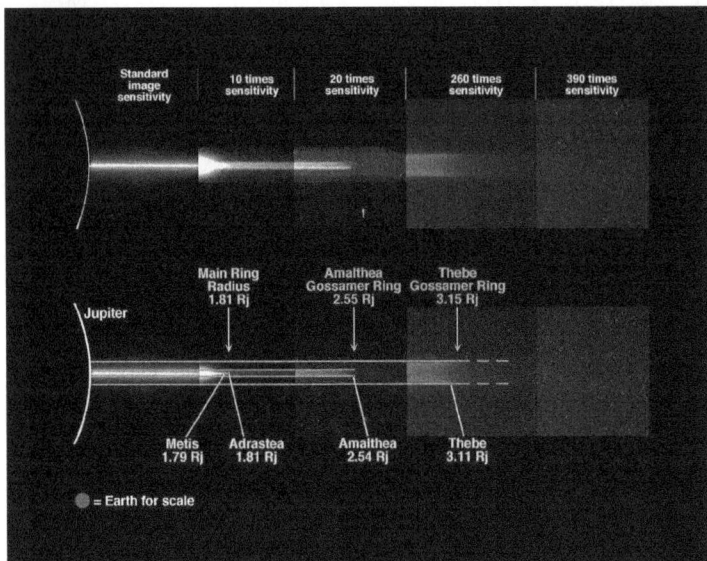

Figure 94: *Mosaic of Jovian ring images with a scheme showing ring and satellite locations*

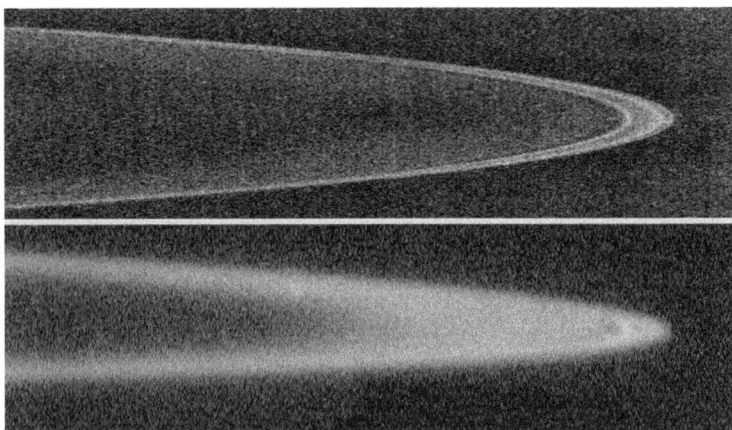

Figure 95: *The upper image shows the main ring in back-scattered light as seen by the New Horizons spacecraft. The fine structure of its outer part is visible. The lower image shows the main ring in forward-scattered light demonstrating its lack of any structure except the Metis notch.*

128000 km, which lacks any visible structure like in the forward-scattering geometry. The Metis notch serves as their boundary. The fine structure of the main ring was discovered in data from the *Galileo* orbiter and is clearly visible in back-scattered images obtained from *New Horizons* in February–March 2007. The early observations by Hubble Space Telescope (HST), Keck and the *Cassini* spacecraft failed to detect it, probably due to insufficient spatial resolution. However the fine structure was observed by the Keck telescope using adaptive optics in 2002–2003.

Observed in back-scattered light the main ring appears to be razor thin, extending in the vertical direction no more than 30 km. In the side scatter geometry the ring thickness is 80–160 km, increasing somewhat in the direction of Jupiter. The ring appears to be much thicker in the forward-scattered light—about 300 km. One of the discoveries of the *Galileo* orbiter was the bloom of the main ring—a faint, relatively thick (about 600 km) cloud of material which surrounds its inner part. The bloom grows in thickness towards the inner boundary of the main ring, where it transitions into the halo.

Detailed analysis of the *Galileo* images revealed longitudinal variations of the main ring's brightness unconnected with the viewing geometry. The Galileo images also showed some patchiness in the ring on the scales 500–1000 km.

In February–March 2007 *New Horizons* spacecraft conducted a deep search for new small moons inside the main ring. While no satellites larger than 0.5 km were found, the cameras of the spacecraft detected seven small clumps of ring particles. They orbit just inside the orbit of Adrastea inside a dense ringlet. The conclusion, that they are clumps and not small moons, is based on their azimuthally extended appearance. They subtend 0.1–0.3° along the ring, which correspond to 1000–3000 km. The clumps are divided into two groups of five and two members, respectively. The nature of the clumps is not clear, but their orbits are close to 115:116 and 114:115 resonances with Metis. They may be wavelike structures excited by this interaction.

Spectra and particle size distribution

Spectra of the main ring obtained by the HST, Keck, Galileo and *Cassini* have shown that particles forming it are red, i.e. their albedo is higher at longer wavelengths. The existing spectra span the range 0.5–2.5 μm. No spectral features have been found so far which can be attributed to particular chemical compounds, although the Cassini observations yielded evidence for absorption bands near 0.8 μm and 2.2 μm. The spectra of the main ring are very similar to Adrastea and Amalthea.

The properties of the main ring can be explained by the hypothesis that it contains significant amounts of dust with 0.1–10 μm particle sizes. This explains

Figure 96: *Image of the main ring obtained by Galileo in forward-scattered light. The Metis notch is clearly visible.*

the stronger forward-scattering of light as compared to back-scattering. However, larger bodies are required to explain the strong back-scattering and fine structure in the bright outer part of the main ring.

Analysis of available phase and spectral data leads to a conclusion that the size distribution of small particles in the main ring obeys a power law

$$n(r) = A \times r^{-q}$$

where $n(r)\, dr$ is a number of particles with radii between r and $r + dr$ and A is a normalizing parameter chosen to match the known total light flux from the ring. The parameter q is 2.0 ± 0.2 for particles with $r < 15 \pm 0.3$ μm and $q = 5 \pm 1$ for those with $r > 15 \pm 0.3$ μm. The distribution of large bodies in the mm–km size range is undetermined presently. The light scattering in this model is dominated by particles with r around 15 μm.

The power law mentioned above allows estimation of the optical depth τ of the main ring: $\tau_l = 4.7 \times 10^{-6}$ for the large bodies and $\tau_s = 1.3 \times 10^{-6}$ for the dust. This optical depth means that the total cross section of all particles inside the ring is about 5000 km². [501] The particles in the main ring are expected to have aspherical shapes. The total mass of the dust is estimated to be 10^7–10^9 kg. The mass of large bodies, excluding Metis and Adrastea, is 10^{11}–10^{16} kg. It depends on their maximum size— the upper value corresponds to about 1 km maximum diameter. These masses can be compared with masses of Adrastea, which is about 2×10^{15} kg, Amalthea, about 2×10^{18} kg, and Earth's Moon, 7.4×10^{22} kg.

The presence of two populations of particles in the main ring explains why its appearance depends on the viewing geometry. The dust scatters light preferably in the forward direction and forms a relatively thick homogenous ring bounded by the orbit of Adrastea. In contrast, large particles, which scatter in the back direction, are confined in a number of ringlets between the Metidian and Adrastean orbits.

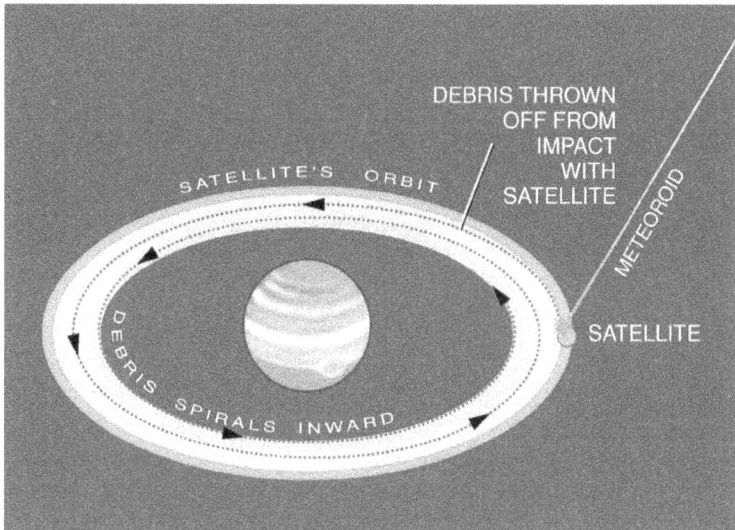

Figure 97: *Formation of Jupiter's rings*

Origin and age

The dust is constantly being removed from the main ring by a combination of Poynting–Robertson drag and electromagnetic forces from the Jovian magnetosphere. Volatile materials, for example ices, evaporate quickly. The lifetime of dust particles in the ring is from 100 to 1000 years, so the dust must be continuously replenished in the collisions between large bodies with sizes from 1 cm to 0.5 km and between the same large bodies and high velocity particles coming from outside the Jovian system. This parent body population is confined to the narrow—about 1000 km—and bright outer part of the main ring, and includes Metis and Adrastea. The largest parent bodies must be less than 0.5 km in size. The upper limit on their size was obtained by *New Horizons* spacecraft. The previous upper limit, obtained from HST and *Cassini* observations, was near 4 km. The dust produced in collisions retains approximately the same orbital elements as the parent bodies and slowly spirals in the direction of Jupiter forming the faint (in back-scattered light) innermost part of the main ring and halo ring. The age of the main ring is currently unknown, but it may be the last remnant of a past population of small bodies near Jupiter.

Figure 98: *False color image of the halo ring obtained by Galileo in forward-scattered light*

Vertical corrugations

Images from the *Galileo* and *New Horizons* space probes show the presence of two sets of spiraling vertical corrugations in the main ring. These waves became more tightly wound over time at the rate expected for differential nodal regression in Jupiter's gravity field. Extrapolating backwards, the more prominent of the two sets of waves appears to have been excited in 1995, around the time of the impact of Comet Shoemaker-Levy 9 with Jupiter, while the smaller set appears to date to the first half of 1990. *Galileo*'s November 1996 observations are consistent with wavelengths of 1920 ± 150 and 630 ± 20 km, and vertical amplitudes of 2.4 ± 0.7 and 0.6 ± 0.2 km, for the larger and smaller sets of waves, respectively. The formation of the larger set of waves can be explained if the ring was impacted by a cloud of particles released by the comet with a total mass on the order of $2\text{–}5 \times 10^{12}$ kg, which would have tilted the ring out of the equatorial plane by 2 km. A similar spiraling wave pattern that tightens over time has been observed by *Cassini* in Saturns's C and D rings.

Halo ring

Appearance and structure

The halo ring is the innermost and the vertically thickest Jovian ring. Its outer edge coincides with the inner boundary of the main ring approximately at the radius 122500 km (1.72 R$_J$). From this radius the ring becomes rapidly thicker towards Jupiter. The true vertical extent of the halo is not known but the presence of its material was detected as high as 10000 km over the ring plane. The inner boundary of the halo is relatively sharp and located at the radius 100000 km (1.4 R$_J$), but some material is present further inward to approximately 92000 km. Thus the width of the halo ring is about 30000 km. Its shape resembles a thick torus without clear internal structure. In contrast to the main ring, the halo's appearance depends only slightly on the viewing geometry.

The halo ring appears brightest in forward-scattered light, in which it was extensively imaged by *Galileo*. While its surface brightness is much less than that of the main ring, its vertically (perpendicular to the ring plane) integrated photon flux is comparable due to its much larger thickness. Despite a claimed vertical extent of more than 20000 km, the halo's brightness is strongly concentrated towards the ring plane and follows a power law of the form $z^{-0.6}$ to $z^{-1.5}$, where z is altitude over the ring plane. The halo's appearance in the back-scattered light, as observed by Keck and HST, is the same. However its total photon flux is several times lower than that of the main ring and is more strongly concentrated near the ring plane than in the forward-scattered light.

The spectral properties of the halo ring are different from the main ring. The flux distribution in the range 0.5–2.5 µm is flatter than in the main ring; the halo is not red and may even be blue.

Origin of the halo ring

The optical properties of the halo ring can be explained by the hypothesis that it comprises only dust with particle sizes less than 15 µm. Parts of the halo located far from the ring plane may consist of submicrometre dust. This dusty composition explains the much stronger forward-scattering, bluer colors and lack of visible structure in the halo. The dust probably originates in the main ring, a claim supported by the fact that the halo's optical depth $\tau_s \sim 10^{-6}$ is comparable with that of the dust in the main ring. The large thickness of the halo can be attributed to the excitation of orbital inclinations and eccentricities of dust particles by the electromagnetic forces in the Jovian magnetosphere. The outer boundary of the halo ring coincides with location of a strong 3:2 Lorentz resonance.[502] As Poynting–Robertson drag causes particles to slowly drift towards Jupiter, their orbital inclinations are excited while passing through it. The bloom of the main ring may be a beginning of the halo. The halo ring's inner boundary is not far from the strongest 2:1 Lorentz resonance. In this resonance the excitation is probably very significant, forcing particles to plunge into the Jovian atmosphere thus defining a sharp inner boundary. Being derived from the main ring, the halo has the same age.

Figure 99: *Image of the gossamer rings obtained by Galileo in forward-scattered light*

Gossamer rings

Amalthea gossamer ring

The Amalthea gossamer ring is a very faint structure with a rectangular cross
section, stretching from the orbit of Amalthea at 182000 km (2.54 R_J) to about
129000 km (1.80 R_J). Its inner boundary is not clearly defined because of
the presence of the much brighter main ring and halo. The thickness of the
ring is approximately 2300 km near the orbit of Amalthea and slightly decreases in the direction of Jupiter.[503] The Amalthea gossamer ring is actually
the brightest near its top and bottom edges and becomes gradually brighter
towards Jupiter; one of the edges is often brighter than another. The outer
boundary of the ring is relatively steep; the ring's brightness drops abruptly
just inward of the orbit of Amalthea, although it may have a small extension
beyond the orbit of the satellite ending near 4:3 resonance with Thebe. In
forward-scattered light the ring appears to be about 30 times fainter than the
main ring. In back-scattered light it has been detected only by the Keck telescope and the ACS (Advanced Camera for Surveys) on HST. Back-scattering
images show additional structure in the ring: a peak in the brightness just inside the Amalthean orbit and confined to the top or bottom edge of the ring.

In 2002–2003 Galileo spacecraft had two passes through the gossamer rings.
During them its dust counter detected dust particles in the size range 0.2–5 μm.
In addition, the Galileo spacecraft's star scanner detected small, discrete bodies
(< 1 km) near Amalthea. These may represent collisional debris generated
from impacts with this satellite.

The detection of the Amalthea gossamer ring from the ground, in *Galileo* images and the direct dust measurements have allowed the determination of the particle size distribution, which appears to follow the same power law as the dust in the main ring with $q=2 \pm 0.5$. The optical depth of this ring is about 10^{-7}, which is an order of magnitude lower than that of the main ring, but the total mass of the dust (10^7-10^9 kg) is comparable.

Thebe gossamer ring

The Thebe gossamer ring is the faintest Jovian ring. It appears as a very faint structure with a rectangular cross section, stretching from the Thebean orbit at 226000 km (3.11 R_J) to about 129000 km (1.80 R_J;). Its inner boundary is not clearly defined because of the presence of the much brighter main ring and halo. The thickness of the ring is approximately 8400 km near the orbit of Thebe and slightly decreases in the direction of the planet. The Thebe gossamer ring is brightest near its top and bottom edges and gradually becomes brighter towards Jupiter—much like the Amalthea ring. The outer boundary of the ring is not especially steep, stretching over 15000 km. There is a barely visible continuation of the ring beyond the orbit of Thebe, extending up to 280000 km (3.75 R_J) and called the Thebe Extension. In forward-scattered light the ring appears to be about 3 times fainter than the Amalthea gossamer ring. In back-scattered light it has been detected only by the Keck telescope. Back-scattering images show a peak of brightness just inside the orbit of Thebe. In 2002–2003 the dust counter of the Galileo spacecraft detected dust particles in the size range 0.2–5 μm—similar to those in the Amalthea ring—and confirmed the results obtained from imaging.

The optical depth of the Thebe gossamer ring is about 3×10^{-8}, which is three times lower than the Amalthea gossamer ring, but the total mass of the dust is the same—about 10^7-10^9 kg. However the particle size distribution of the dust is somewhat shallower than in the Amalthea ring. It follows a power law with q < 2. In the Thebe extension the parameter q may be even smaller.

Origin of the gossamer rings

The dust in the gossamer rings originates in essentially the same way as that in the main ring and halo. Its sources are the inner Jovian moons Amalthea and Thebe respectively. High velocity impacts by projectiles coming from outside the Jovian system eject dust particles from their surfaces. These particles initially retain the same orbits as their moons but then gradually spiral inward by Poynting–Robertson drag. The thickness of the gossamer rings is determined by vertical excursions of the moons due to their nonzero orbital inclinations. This hypothesis naturally explains almost all observable properties of the rings:

rectangular cross-section, decrease of thickness in the direction of Jupiter and brightening of the top and bottom edges of the rings.

However some properties have so far gone unexplained, like the Thebe Extension, which may be due to unseen bodies outside Thebe's orbit, and structures visible in the back-scattered light. One possible explanation of the Thebe Extension is influence of the electromagnetic forces from the Jovian magnetosphere. When the dust enters the shadow behind Jupiter, it loses its electrical charge fairly quickly. Since the small dust particles partially corotate with the planet, they will move outward during the shadow pass creating an outward extension of the Thebe gossamer ring. The same forces can explain a dip in the particle distribution and ring's brightness, which occurs between the orbits of Amalthea and Thebe.

The peak in the brightness just inside of the Amalthea's orbit and, therefore, the vertical asymmetry the Amalthea gossamer ring may be due to the dust particles trapped at the leading (L_4) and trailing (L_5) Lagrange points of this moon. The particles may also follow horseshoe orbits between the Lagrangian points. The dust may be present at the leading and trailing Lagrange points of Thebe as well. This discovery implies that there are two particle populations in the gossamer rings: one slowly drifts in the direction of Jupiter as described above, while another remains near a source moon trapped in 1:1 resonance with it.

Himalia ring

The small moon Dia, 4 kilometres in diameter, had gone missing since its discovery in 2000. One theory was that it had crashed into the much larger moon Himalia, 170 kilometres in diameter, creating a faint ring. This possible ring appears as a faint streak near Himalia in images from NASA's *New Horizons* mission to Pluto. This suggests that Jupiter sometimes gains and loses small moons through collisions. However, the recovery of Dia in 2010 and 2011 disproves the link between Dia and the Himalia ring, although it is still possible that a different moon may have been involved.

Exploration

The existence of the Jovian rings was inferred from observations of the planetary radiation belts by Pioneer 11 spacecraft in 1975. In 1979 the *Voyager 1* spacecraft obtained a single overexposed image of the ring system. More extensive imaging was conducted by *Voyager 2* in the same year, which allowed rough determination of the ring's structure. The superior quality of the images obtained by the *Galileo* orbiter between 1995 and 2003 greatly extended the

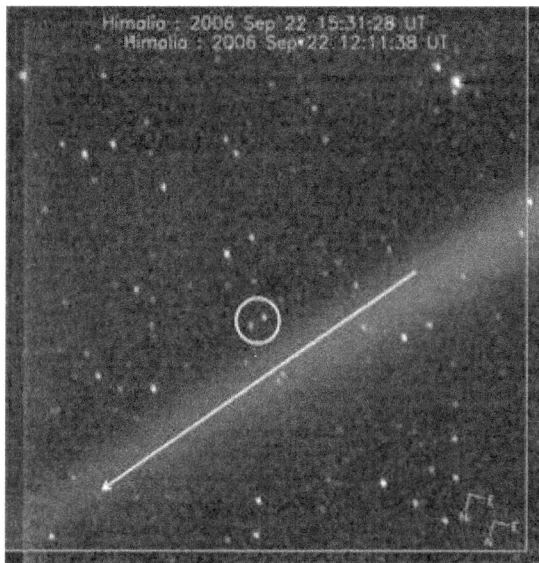

Figure 100: *New Horizons image of possible Himalia ring*

existing knowledge about the Jovian rings. Ground-based observation of the rings by the Keck telescope in 1997 and 2002 and the HST in 1999 revealed the rich structure visible in back-scattered light. Images transmitted by the *New Horizons* spacecraft in February–March 2007 allowed observation of the fine structure in the main ring for the first time. In 2000, the *Cassini* spacecraft en route to Saturn conducted extensive observations of the Jovian ring system. Future missions to the Jovian system will provide additional information about the rings.

Gallery

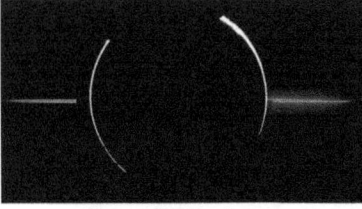

Figure 101: *The ring system as imaged by Galileo*

Figure 102: *The rings as observed from the inside by Juno on 27 August 2016*

External links

 Wikimedia Commons has media related to *Jupiter (rings)*.

- Jupiter Rings Fact Sheet[504]
- Jupiter's Rings[505] by NASA's Solar System Exploration[506]
- NASA Pioneer project page[507]
- NASA Voyager project page[508]
- NASA Galileo project page[509]
- NASA Cassini project space[510]
- New Horizons project page[511]
- Planetary Ring Node: Jupiter's Ring System[512]
- Rings of Jupiter nomenclature[513] from the USGS planetary nomenclature page[514]

<indicator name="featured-star"> ⭐ </indicator>

Appendix

References

[1] This image was taken by the Hubble Space Telescope, using the Wide Field Camera 3, on April 21, 2014. Jupiter's atmosphere and its appearance constantly changes, and hence its current appearance today may not resemble what it was when this image was taken. Depicted in this image, however, are a few features that remain consistent, such as the famous Great Red Spot, featured prominently in the lower right of the image, and the planet's recognizable banded appearance.

[2] //en.wikipedia.org/w/index.php?title=Jupiter&action=edit

[3]

[4] Based on the volume within the level of 1 bar atmospheric pressure

[5] Philosophical Transactions Vol. I http://www.gutenberg.org/files/28758/28758-h/28758-h.htm (1665–1666.). Project Gutenberg. Retrieved on December 22, 2011.

[6] – See section 3.4.

[7] See also the Greek article about the planet.

[8] See for example: "IAUC 2844: Jupiter; 1975h" http://cbat.eps.harvard.edu/iauc/02800/02844. html. International Astronomical Union. October 1, 1975. Retrieved October 24, 2010. That particular word has been in use since at least 1966. See: "Query Results from the Astronomy Database" http://adsabs.harvard.edu/cgi-bin/nph-abs_connect?db_key=AST&text= zenographic%20since%20at%20least%201966. Smithsonian/NASA. Retrieved July 29, 2007.

[9] The Aryabhatiya of Aryabhata https://archive.org/details/The_Aryabhatiya_of_Aryabhata_ Clark_1930 is also available at Internet Archive.

[10] NASA – Pioneer 10 Mission Profile http://quest.nasa.gov/sso/cool/pioneer10/mission/ . NASA. Retrieved on December 22, 2011.

[11] NASA – Glenn Research Center http://www.nasa.gov/centers/glenn/about/history/pioneer. html. NASA. Retrieved on December 22, 2011.

[12] Fortescue, Peter W.; Stark, John and Swinerd, Graham Spacecraft systems engineering, 3rd ed., John Wiley and Sons, 2003, p. 150.

[13] New approach for L-class mission candidates http://sci.esa.int/science-e/www/object/index. cfm?fobjectid=48661, ESA, April 19, 2011

[14] //www.worldcat.org/issn/0027-9358

[15] //www.worldcat.org/oclc/643483454

[16] http://www.dtic.mil/docs/citations/ADA471037

[17] http://adsabs.harvard.edu/abs/2007CeMDA..98..155S

[18] //doi.org/10.1007/s10569-007-9072-y

[19] https://books.google.com/books?id=stFpBgAAQBAJ&pg=PA250

[20] http://www.vias.org/spacetrip/jupiter_1.html

[21] http://www.vias.org/spacetrip/index.html

[22] http://orbitsimulator.com/gravity/articles/joviansystem.html

[23] https://archive.is/20121210054932/http://skytonight.com/observing/objects/planets/3307071. html?page=1&c=y

[24] http://skytonight.com/observing/objects/planets/3307071.html?page=1&c=y

[25] http://news.bbc.co.uk/2/hi/in_pictures/6614557.stm

[26] http://www.astronomycast.com/2007/10/episode-56-jupiter/

[27] https://web.archive.org/web/20111020171302/http://science1.nasa.gov/science-news/science-at-nasa/2007/01may_fantasticflyby/

[28] http://science1.nasa.gov/science-news/science-at-nasa/2007/01may_fantasticflyby/

[29] http://www.psrd.hawaii.edu/Archive/Archive-Jupiter.html

[30] https://www.flickr.com/photos/136797589@N04/albums/72157685871712710

[31] https://www.youtube.com/watch?v=Us6EXc5Hyng

[32] http://www.sixtysymbols.com/videos/jupiter.htm

[33] http://solarsystem.nasa.gov/planets/profile.cfm?Object=Jupiter&Display=Facts

[34]http://digitalcollections.ucsc.edu/cdm/search/collection/p265101coll10/searchterm/Jupiter%20(planet)/order/title

[35]Some of the values in this table are nominal values, derived from *Numerical Standards for Fundamental Astronomy* and rounded using appropriate attention to significant figures, as recommended by the IAU Resolution B3.

[36]Atreya Mahaffy Niemann et al. 2003.

[37]Atreya & Wong 2005.

[38]

[39]Yelle (2004), p. 1

[40]Miller Aylward et al. 2005.

[41]

[42]Atreya Wong Baines et al. 2005.

[43]Atreya Wong Owen et al. 1999.

[44]West *et al.* (2004), pp. 9–10, 20–23

[45]

[46]Ingersoll (2004), p. 8

[47]Yelle (2004), pp. 1–12

[48]

[49]Yelle (2004), pp. 22–27

[50]Bhardwaj & Gladstone 2000, pp. 299–302.

[51]Encrenaz 2003.

[52]Kunde *et al.* (2004)

[53]

[54]Rogers (1995), p. 81.

[55]

[56]

[57]

[58]

[59]Graney (2010)

[60]Rogers (1995), pp. 85, 91–4.

[61]Rogers (1995), pp. 101–105.

[62]Rogers (1995), pp.113–117.

[63]Rogers (1995), pp. 125–130.

[64]Rogers (1995), pp. 133, 145–147.

[65]Rogers (1995), p. 133.

[66]Beebe (1997), p. 24.

[67]Rogers (1995), pp. 159–160

[68]Rogers (1995), pp. 219–221, 223, 228–229.

[69]Rogers (1995), p. 235.

[70]Rogers *et al.* (2003)

[71]Rogers and Metig (2001)

[72]Ridpath (1998)

[73]

[74]Vasavada (2005), pp. 1943–1945

[75]Heimpel *et al.* (2005)

[76]See, e. g., Ingersoll *et al.* (1969)

[77]Ingersoll (2004), pp. 16–17

[78]Ingersoll (2004), pp. 14–15

[79]

[80]Vasavada (2005), p. 1949

[81]Vasavada (2005), pp. 1945–1947

[82]Vasavada (2005), pp. 1962–1966

[83]Vasavada (2005), p. 1966

[84]Busse (1976)

[85]Vasavada (2005), pp. 1966–1972

[86]Vasavada (2005), p. 1970

[87] Low (1966)

[88] Pearl Conrath et al. 1990, pp. 12, 26.

[89] Ingersoll (2004), pp. 11, 17–18

[90]

[91] Vasavada (2005), p. 1978

[92] Vasavada (2005), p. 1977

[93]

[94]

[95] Vasavada (2005), p. 1975

[96]

[97]

[98] Vasavada (2005), p. 1979

[99] Graney (2010), p. 266.

[100] Smith *et al.* (1979), p. 954.

[101] Irwin, 2003, p. 171

[102] Beatty (2002)

[103] Rogers (1995), p. 191.

[104] Rogers (1995), pp. 194–196.

[105] Beebe (1997), p. 35.

[106] Rogers (1995), p. 195.

[107] Fletcher (2010), p. 306

[108] Reese and Gordon (1966)

[109] Rogers (1995), 192–193.

[110] Stone (1974)

[111] Rogers (1995), pp. 48, 193.

[112] Rogers (1995), p. 193.

[113] Beebe (1997), pp. 38–41.

[114] Is Jupiter's Great Red Spot a Sunburn? https://science.nasa.gov/science-news/science-at-nasa/2014/28nov_sunburn/ NASA.com November 28, 2014

[115] Jupiter's Red Spot is Likely a Sunburn, Not a Blush http://www.nasa.gov/jpl/cassini/jupiters-red-spot-is-likely-a-sunburn-not-a-blush/ NASA.com, November 11, 2014

[116] Hammel *et al.* (1995), p. 1740

[117] Sanchez-Lavega *et al.* (2001)

[118] Rogers (1995), p. 223.

[119] Go *et al.* (2006)

[120] Vasavada (2005), pp. 1982, 1985–1987

[121] Sanchez-Lavega *et al.* (2008), pp. 437–438

[122] Vasavada (2005), pp. 1983–1985

[123] Baines Simon-Miller et al. 2007, p. 226.

[124] McKim (1997)

[125] Ingersoll (2004), p. 2

[126] Noll (1995), p. 1307

[127] Rogers (1995), p. 6.

[128] Rogers (2008), pp.111–112

[129] Rogers (1995), p. 188

[130] Hockey, 1999, pp. 40–41.

[131] Smith *et al.* (1979), pp. 951–972.

[132] Rogers (1995), pp. 224–5.

[133] Rogers (1995), p. 226–227.

[134] Rogers (1995), p. 226.

[135] Rogers (1995), p. 225.

[136] Beebe (1997), p. 43.

[137] http://adsabs.harvard.edu/abs/1999P&SS...47.1243A

[138] //doi.org/10.1016/S0032-0633%2899%2900047-1

[139] //www.worldcat.org/issn/0032-0633

[140] //www.ncbi.nlm.nih.gov/pubmed/11543193

[141] http://adsabs.harvard.edu/abs/2003P&SS...51..105A
[142] //doi.org/10.1016/S0032-0633%2802%2900144-7
[143] http://www-personal.umich.edu/~atreya/Chapters/2005_JovianCloud_Multiprobes.pdf
[144] http://adsabs.harvard.edu/abs/2005SSRv..116..121A
[145] //doi.org/10.1007/s11214-005-1951-5
[146] http://www-personal.umich.edu/~atreya/Articles/2005_Jupiters_Ammonia.pdf
[147] http://adsabs.harvard.edu/abs/2005P&SS...53..498A
[148] //doi.org/10.1016/j.pss.2004.04.002
[149] http://adsabs.harvard.edu/abs/2007Sci...318..226B
[150] //doi.org/10.1126/science.1147912
[151] //www.ncbi.nlm.nih.gov/pubmed/17932285
[152] http://www.saburchill.com/HOS/astronomy/034.html
[153] //www.worldcat.org/oclc/224014042
[154] http://www.bu.edu/csp/uv/cp-aeronomy/Bhardwaj_Gladstone_RG_2000.pdf
[155] http://adsabs.harvard.edu/abs/2000RvGeo..38..295B
[156] //doi.org/10.1029/1998RG000046
[157] http://adsabs.harvard.edu/abs/1976Icar...29..255B
[158] //doi.org/10.1016/0019-1035%2876%2990053-1
[159] http://adsabs.harvard.edu/abs/2003P&SS...51...89E
[160] //doi.org/10.1016/S0032-0633%2802%2900145-9
[161] http://www.eso.org/public/archives/releases/sciencepapers/eso1010/eso1010.pdf
[162] http://adsabs.harvard.edu/abs/2010Icar..208..306F
[163] //doi.org/10.1016/j.icarus.2010.01.005
[164] http://adsabs.harvard.edu/abs/2006DPS....38.1102G
[165] //arxiv.org/abs/1008.0566
[166] http://adsabs.harvard.edu/abs/2010BaltA..19..265G
[167] //arxiv.org/abs/astro-ph/9907402
[168] http://adsabs.harvard.edu/abs/1999P&SS...47.1183G
[169] //doi.org/10.1016/S0032-0633%2899%2900043-4
[170] http://adsabs.harvard.edu/abs/1995Sci...268.1740H
[171] //doi.org/10.1126/science.268.5218.1740
[172] //www.ncbi.nlm.nih.gov/pubmed/17834994
[173] http://adsabs.harvard.edu/abs/2005Natur.438..193H
[174] //doi.org/10.1038/nature04208
[175] //www.ncbi.nlm.nih.gov/pubmed/16281029
[176] //www.worldcat.org/oclc/39733730
[177] http://www.lpl.arizona.edu/~showman/publications/ingersolletal-2004.pdf
[178] http://adsabs.harvard.edu/abs/1969JAtS...26..981I
[179] //doi.org/10.1175/1520-0469%281969%29026%3C0981%3ADOJCB%3E2.0.CO%3B2
[180] http://adsabs.harvard.edu/abs/2004Sci...305.1582K
[181] //doi.org/10.1126/science.1100240
[182] //www.ncbi.nlm.nih.gov/pubmed/15319491
[183] http://adsabs.harvard.edu/abs/1966AJ.....71R.391L
[184] //doi.org/10.1086/110110
[185] http://adsabs.harvard.edu/abs/1997JBAA..107..239M
[186] http://adsabs.harvard.edu/abs/2005SSRv..116..319M
[187] //doi.org/10.1007/s11214-005-1960-4
[188] http://adsabs.harvard.edu/abs/1995Sci...267.1307N
[189] //doi.org/10.1126/science.7871428
[190] //www.ncbi.nlm.nih.gov/pubmed/7871428
[191] http://adsabs.harvard.edu/abs/1990Icar...84...12P
[192] //doi.org/10.1016/0019-1035%2890%2990155-3
[193] //www.worldcat.org/issn/0019-1035
[194] http://adsabs.harvard.edu/abs/1966Icar....5..266R
[195] //doi.org/10.1016/0019-1035%2866%2990036-4
[196] //www.worldcat.org/oclc/219591510

[197] http://www.britastro.org/jupiter/JBAA111_Jup98.pdf
[198] http://adsabs.harvard.edu/abs/2001JBAA..111..321R
[199] http://www.britastro.org/jupiter/JBAA113-Jup9900_II-IR.pdf
[200] http://adsabs.harvard.edu/abs/2003JBAA..113..136R
[201] http://www.britastro.org/jupiter/JBAA-118-1_GRS-paper.pdf
[202] http://adsabs.harvard.edu/abs/2008JBAA..118...14R
[203] http://adsabs.harvard.edu/abs/2001Icar..149..491S
[204] //doi.org/10.1006/icar.2000.6548
[205] http://adsabs.harvard.edu/abs/2008Natur.451..437S
[206] //doi.org/10.1038/nature06533
[207] //www.ncbi.nlm.nih.gov/pubmed/18216848
[208] http://adsabs.harvard.edu/abs/1998JGR...10322857S
[209] //doi.org/10.1029/98JE01766
[210] http://adsabs.harvard.edu/abs/1979Sci...204..951S
[211] //doi.org/10.1126/science.204.4396.951
[212] //www.ncbi.nlm.nih.gov/pubmed/17800430
[213] http://journals.ametsoc.org/doi/abs/10.1175/1520-0469%281974%29031%3C1471%3AOJROR%3E2.0.CO%3B2
[214] http://adsabs.harvard.edu/abs/1974JAtS...31.1471S
[215] //doi.org/10.1175/1520-0469%281974%29031%3C1471%3AOJROR%3E2.0.CO%3B2
[216] http://adsabs.harvard.edu/abs/2005RPPh...68.1935V
[217] //doi.org/10.1088/0034-4885/68/8/R06
[218] http://www.gps.caltech.edu/~ulyana/www_papers/west_Ch5_us.pdf
[219] http://www.lpl.arizona.edu/~yelle/eprints/Yelle04c.pdf
[220] //www.worldcat.org/oclc/39464951
[221] //www.worldcat.org/oclc/8318939
[222] http://www.berkeley.edu/news/media/releases/2004/04/21_jupiter.shtml
[223] https://web.archive.org/web/20070609214744/http://www.berkeley.edu/news/media/releases/2004/04/21_jupiter.shtml
[224] http://adsabs.harvard.edu/abs/2003Icar..162...74Y
[225] //doi.org/10.1016/S0019-1035%2802%2900060-X
[226] http://www.gfdl.noaa.gov/bibliography/related_files/gw7501.pdf
[227] http://adsabs.harvard.edu/abs/1975Natur.257..778W
[228] //doi.org/10.1038/257778a0
[229] http://www.gfdl.noaa.gov/bibliography/related_files/gw7801.pdf
[230] http://adsabs.harvard.edu/abs/1978JAtS...35.1399W
[231] //doi.org/10.1175/1520-0469%281978%29035%3C1399%3APCBROJ%3E2.0.CO%3B2
[232] http://www.gfdl.noaa.gov/bibliography/related_files/gw8502.pdf
[233] http://adsabs.harvard.edu/abs/1985AdGeo..28..381W
[234] //doi.org/10.1016/S0065-2687%2808%2960231-9
[235] http://www.gfdl.noaa.gov/bibliography/related_files/gw9701.pdf
[236] http://adsabs.harvard.edu/abs/1997JGR...102.9303W
[237] //doi.org/10.1029/97JE00520
[238] http://www.gfdl.noaa.gov/bibliography/related_files/gw9601.pdf
[239] http://adsabs.harvard.edu/abs/1996JAtS...53.2685W
[240] //doi.org/10.1175/1520-0469%281996%29053%3C2685%3AJDPVSS%3E2.0.CO%3B2
[241] http://www.gfdl.noaa.gov/bibliography/related_files/gpw0201.pdf
[242] http://adsabs.harvard.edu/abs/2002JAtS...59.1356W
[243] //doi.org/10.1175/1520-0469%282002%29059%3C1356%3AJDPITG%3E2.0.CO%3B2
[244] http://www.gfdl.noaa.gov/bibliography/related_files/gw0301.pdf
[245] http://adsabs.harvard.edu/abs/2003JAtS...60.1270W
[246] //doi.org/10.1175/1520-0469%282003%29060%3C1270%3AJDPIMM%3E2.0.CO%3B2
[247] http://www.gfdl.noaa.gov/bibliography/related_files/gw0304.pdf
[248] http://www.gfdl.noaa.gov/bibliography/related_files/gw0303.pdf
[249] http://adsabs.harvard.edu/abs/2003JAtS...60.2136W
[250] //doi.org/10.1175/1520-0469%282003%29060%3C2136%3ABIAES%3E2.0.CO%3B2

[251] http://www.gfdl.noaa.gov/bibliography/related_files/gw0302.pdf

[252] //doi.org/10.2151/jmsj.81.439

[253] http://www.gfdl.noaa.gov/bibliography/related_files/gw0601.pdf

[254] http://adsabs.harvard.edu/abs/2006JAtS...63.1548W

[255] //doi.org/10.1175/JAS3711.1

[256] http://www.planetary.org/blogs/guest-blogs/2017/201170509-journey-to-jupiter.html

[257] https://www.youtube.com/watch?v=YZc1Y662jtk

[258] https://www.britannica.com/place/Jupiter-planet/Basic-astronomical-data#toc54253

[259] Russel, 1993, p. 694

[260] Blanc, 2005, p. 238 (Table III)

[261] Khurana, 2004, pp. 12–13

[262] The north and south poles of the Earth's dipole should not be confused with Earth's North magnetic pole and South magnetic pole, which lie in the northern and southern hemispheres, respectively.

[263]

[264] The magnetic moment is proportional to the product of the equatorial field strength and cube of Jupiter's radius, which is 11 times larger than that of the Earth.

[265] For instance, the azimuthal orientation of the dipole changed by less than $0.01°$.

[266]

[267] Russell, 2001, pp. 1015–1016

[268] Krupp, 2004, pp. 15–16

[269] Russel, 1993, pp. 725–727

[270]

[271] Khurana, 2004, pp. 6–7

[272]

[273]

[274] Krupp, 2004, pp. 3–4

[275] Krupp, 2004, pp. 1–3

[276]

[277] Khurana, 2004, pp. 10–12

[278]

[279] Khurana, 2004, pp. 20–21

[280]

[281] Kivelson, 2005, pp. 315–316

[282]

[283] Blanc, 2005, pp. 250–253

[284] Cowley, 2001, pp. 1069–76

[285] The direct current in the Jovian magnetosphere is not to be confused with the direct current used in electrical circuits. The latter is the opposite of the alternating current.

[286] Blanc, 2005, pp. 254–261

[287] Cowley, 2001, pp. 1083–87

[288] Russell, 2008

[289] Krupp, 2004, pp. 7–9

[290]

[291] Khurana, 2004, pp. 18–19

[292]

[293] Nichols, 2006, pp. 393–394

[294] The Jovian ionosphere is another significant source of protons.

[295] Krupp, 2004, pp. 18–19

[296] Nichols, 2006, pp. 404–405

[297]

[298] Bhardwaj, 2000, pp. 311–316

[299]

[300] Cowley, 2003, pp. 49–53

[301] Bhardwaj, 2000, pp. 316–319

[302] Bhardwaj, 2000, pp. 306–311

[303] Bhardwaj, 2000, p. 296

[304] Miller Aylward et al. 2005, pp. 335–339.

[305] Bhardwaj, 2000, Tables 2 and 5

[306] Callisto may have a spot as well; however, it would be unobservable because it would coincide with the main auroral oval.Russel, 1993, p. 694

[307] Clarke, 2002

[308] Blanc, 2005, pp. 277–283

[309] Palier, 2001, pp. 1170–71

[310]

[311] Zarka, 1998, pp. 20,160–168

[312] The non-Io-DAM is much weaker than the Io-DAM, and is the high-frequency tail of the HOM emissions.

[313] Zarka, 1998, pp. 20, 173–181

[314] Hill, 1995

[315] Santos-Costa, 2001

[316]

[317] Zarka, 2005, pp. 384–385

[318] Krupp, 2004, pp. 17–18

[319] Johnson, 2004, pp. 1–2

[320]

[321] Burns, 2004, pp. 12–14

[322] Burns, 2004, pp. 10–11

[323] A Lorentz resonance is one that exists between a particle's orbital speed and the rotation period of a planet's magnetosphere. If the ratio of their angular frequencies is $m:n$ (a rational number) then scientists call it an $m:n$ Lorentz resonance. So, in the case of a 3:2 resonance, a particle at a distance of about 1.71 R_J from Jupiter makes three revolutions around the planet, while the planet's magnetic field makes two revolutions.Blanc, 2005, p. 238 (Table III)

[324] Burns, 2004, pp. 17–19

[325]

[326]

[327]

[328] Cooper, 2001, pp. 137,139

[329]

[330]

[331] Williams, 1998, p. 1

[332] Cooper, 2001, pp. 154–156

[333] Hibbitts, 2000, p. 1

[334] Burke, 1955

[335]

[336] Drake, 1959

[337]

[338]

[339] Pioneer 10 carried a helium vector magnetometer, which measured the magnetic field of Jupiter directly. The spacecraft also made observations of plasma and energetic particles.

[340]

[341] Fieseler, 2002

[342]

[343] Troutman, 2003

[344] "Jupiter's Magnetosphere and Aurorae Observed by the Juno Spacecraft During its First Polar Orbits." (PDF). J. E.P. Connerney, A. Adriani, F. Allegrini, et al. *Science Journal*, 26 May 2017: Vol. 356, Issue 6340, pp. 826-832.

[345] http://www.bu.edu/csp/uv/cp-aeronomy/Bhardwaj_Gladstone_RG_2000.pdf

[346] http://adsabs.harvard.edu/abs/2000RvGeo..38..295B

[347] //doi.org/10.1029/1998RG000046

[348] http://adsabs.harvard.edu/abs/2005SSRv..116..227B

[349] //doi.org/10.1007/s11214-005-1958-y

350 http://www.nature.com/nature/journal/v415/n6875/full/415987a.html
351 http://adsabs.harvard.edu/abs/2002Natur.415..987B
352 //doi.org/10.1038/415987a
353 //www.ncbi.nlm.nih.gov/pubmed/11875557
354 http://adsabs.harvard.edu/abs/1955JGR....60..213B
355 //doi.org/10.1029/JZ060i002p00213
356 http://www.astro.umd.edu/~hamilton/research/preprints/BurSimSho03.pdf
357 http://adsabs.harvard.edu/abs/2004jpsm.book..241B
358 http://www2.iap.fr/users/lotfi/jupiter.pdf
359 //doi.org/10.1038/415997a
360 //www.ncbi.nlm.nih.gov/pubmed/11875560
361 https://web.archive.org/web/20090225131107/http://icymoons.com/europaclass/Cooper_
    gllsat_irrad.pdf
362 http://adsabs.harvard.edu/abs/2001Icar..149..133C
363 //doi.org/10.1006/icar.2000.6498
364 http://icymoons.com/europaclass/Cooper_gllsat_irrad.pdf
365 http://adsabs.harvard.edu/abs/2001P&SS...49.1067C
366 //doi.org/10.1016/S0032-0633%2800%2900167-7
367 http://adsabs.harvard.edu/abs/2003P&SS...51...31C
368 //doi.org/10.1016/S0032-0633%2802%2900130-7
369 http://adsabs.harvard.edu/abs/1959AJ.....64S.329D
370 //doi.org/10.1086/108047
371 http://wwwastro.msfc.nasa.gov/research/papers/icarus.pdf
372 http://adsabs.harvard.edu/abs/2005Icar..178..417E
373 //doi.org/10.1016/j.icarus.2005.06.006
374 https://web.archive.org/web/20110719204111/http://trs-new.jpl.nasa.gov/dspace/bitstream/
    2014/11661/1/02-0220.pdf
375 http://adsabs.harvard.edu/abs/2002ITNS...49.2739F
376 //doi.org/10.1109/TNS.2002.805386
377 http://trs-new.jpl.nasa.gov/dspace/bitstream/2014/11661/1/02-0220.pdf
378 https://web.archive.org/web/19970501040601/http://www.agu.org/sci_soc/hill.html
379 http://adsabs.harvard.edu/abs/1995EOSTr..76..313H
380 //doi.org/10.1029/95EO00190
381 http://www.agu.org/sci_soc/hill.html
382 http://adsabs.harvard.edu/abs/2000JGR...10522541H
383 //doi.org/10.1029/1999JE001101
384 http://people.virginia.edu/~rej/papers04/chap20.pdf
385 http://www.igpp.ucla.edu/people/mkivelson/Publications/279-Ch24.pdf
386 http://www.igpp.ucla.edu/people/mkivelson/Publications/285-SSR11629905.pdf
387 http://adsabs.harvard.edu/abs/2005SSRv..116..299K
388 //doi.org/10.1007/s11214-005-1959-x
389 http://www.igpp.ucla.edu/people/mkivelson/Publications/277-Ch21.pdf
390 http://www.igpp.ucla.edu/people/mkivelson/Publications/280-Ch25.pdf
391 http://adsabs.harvard.edu/abs/2007Sci...318..216K
392 //doi.org/10.1126/science.1150448
393 //www.ncbi.nlm.nih.gov/pubmed/17932281
394 http://adsabs.harvard.edu/abs/2005SSRv..116..319M
395 //doi.org/10.1007/s11214-005-1960-4
396 http://www.ann-geophys.net/24/393/2006/angeo-24-393-2006.html
397 http://adsabs.harvard.edu/abs/2006AnGeo..24..393N
398 //doi.org/10.5194/angeo-24-393-2006
399 http://adsabs.harvard.edu/abs/2001P&SS...49.1159P
400 //doi.org/10.1016/S0032-0633%2801%2900023-X
401 http://www.iop.org/EJ/article/0034-4885/56/6/001/rp930601.pdf
402 http://adsabs.harvard.edu/abs/1993RPPh...56..687R
403 //doi.org/10.1088/0034-4885/56/6/001

[404] http://adsabs.harvard.edu/abs/2001P&SS...49.1005R
[405] //doi.org/10.1016/S0032-0633%2801%2900017-4
[406] http://www-ssc.igpp.ucla.edu/personnel/russell/papers/magnetospheres_jupiter_saturn.pdf
[407] http://adsabs.harvard.edu/abs/2008AdSpR..41.1310R
[408] //doi.org/10.1016/j.asr.2007.07.037
[409] http://adsabs.harvard.edu/abs/2001P&SS...49..303S
[410] //doi.org/10.1016/S0032-0633%2800%2900151-3
[411] http://adsabs.harvard.edu/abs/1974JGR....79.3501S
[412] //doi.org/10.1029/JA079i025p03501
[413] //doi.org/10.1063/1.1541373
[414] http://adsabs.harvard.edu/abs/1998JGR...10317523W
[415] //doi.org/10.1029/98JA01370
[416] http://adsabs.harvard.edu/abs/1998JGR...10320159Z
[417] //doi.org/10.1029/98JE01323
[418] http://adsabs.harvard.edu/abs/2005SSRv..116..371Z
[419] //doi.org/10.1007/s11214-005-1962-2
[420] http://adsabs.harvard.edu/abs/1969ARA&A...7..577C
[421] //doi.org/10.1146/annurev.aa.07.090169.003045
[422] http://adsabs.harvard.edu/abs/2001P&SS...49.1115E
[423] //doi.org/10.1016/S0032-0633%2800%2900164-1
[424] http://www.nature.com/nature/journal/v415/n6875/full/4151000a.html
[425] http://adsabs.harvard.edu/abs/2002Natur.415.1000G
[426] //doi.org/10.1038/4151000a
[427] //www.ncbi.nlm.nih.gov/pubmed/11875561
[428] http://www.igpp.ucla.edu/people/mkivelson/Publications/262-2001JA000251.pdf
[429] http://adsabs.harvard.edu/abs/2002JGRA..107.1116K
[430] //doi.org/10.1029/2001JA000251
[431] http://www.igpp.ucla.edu/people/mkivelson/Publications/287-ASR362077.pdf
[432] http://adsabs.harvard.edu/abs/2005AdSpR..36.2077K
[433] //doi.org/10.1016/j.asr.2005.05.104
[434] http://www.igpp.ucla.edu/people/mkivelson/Publications/270-PSS51891.pdf
[435] http://adsabs.harvard.edu/abs/2003P&SS...51..891K
[436] //doi.org/10.1016/S0032-0633%2803%2900075-8
[437] http://adsabs.harvard.edu/abs/2007Sci...318..217M
[438] //doi.org/10.1126/science.1147393
[439] //www.ncbi.nlm.nih.gov/pubmed/17932282
[440] http://adsabs.harvard.edu/abs/2001P&SS...49..275M
[441] //doi.org/10.1016/S0032-0633%2800%2900148-3
[442] http://www.igpp.ucla.edu/people/mkivelson/Publications/245-2001GL012917.pdf
[443] http://adsabs.harvard.edu/abs/2001GeoRL..28.1911R
[444] //doi.org/10.1029/2001GL012917
[445] http://adsabs.harvard.edu/abs/2001P&SS...49.1137Z
[446] //doi.org/10.1016/S0032-0633%2801%2900021-6
[447] //en.wikipedia.org/w/index.php?title=Exploration_of_Jupiter&action=edit
[448] Fischer, 1999, p. 44
[449] CRC Handbook of Chemistry and Physics, 64th EDITION, (C) 1983, page F-141
[450] (discovery)
[451] NASA Selects New Frontiers Concept Study: Juno Mission to Jupiter I Jupiter Today – Your Daily Source of Jupiter News https://web.archive.org/web/20110713125816/http://www.jupitertoday.com/news/viewpr.rss.html?pid=16990
[452] *Executive Survey* (Visions and Voyages for Planetary Science 2013 – 2022) https://solarsystem.nasa.gov/multimedia/downloads/Decadal_Exec_Summary.pdf
[453] Artemis Society International http://www.asi.org/ official website
[454] Robert Zubrin, *Entering Space: Creating a Spacefaring Civilization*, section: Settling the Outer Solar System: The Sources of Power, pp. 158–160, Tarcher/Putnam, 1999,

[455] Jeffrey Van Cleve (Cornell University) et al., "Helium-3 Mining Aerostats in the Atmosphere of Uranus" http://www.mines.edu/research/srr/2001abstracts/vancleve.pdf , Abstract for Space Resources Roundtable, accessed May 10, 2006

[456] Robert Zubrin, *Entering Space: Creating a Spacefaring Civilization*, section: Colonizing the Jovian System, pp. 166–170, Tarcher/Putnam, 1999, .

[457] http://nssdc.gsfc.nasa.gov/planetary/chronology.html

[458] http://nssdc.gsfc.nasa.gov/planetary/planets/jupiterpage.html

[459] For comparison, the area of a sphere with diameter 250 km is about the area of Senegal and comparable to the area of Belarus, Syria and Uruguay. The area of a sphere with diameter 5 km is about the area of Guernsey and somewhat more than the area of San Marino. (But note that these smaller moons are not spherical.)

[460] Gazetteer of Planetary Nomenclature https://planetarynames.wr.usgs.gov/Page/Planets Planet and Satellite Names and Discoverers
International Astronomical Union (IAU)

[461] Jupiter Mass of 1.8986 $\times$ 10^{27} kg / Mass of Galilean moons http://ssd.jpl.nasa.gov/?sat_phys_par 3.93 $\times$ 10^{23} kg = 4,828

[462] Order refers to the position among other moons with respect to their average distance from Jupiter.

[463] Label refers to the Roman numeral attributed to each moon in order of their naming.

[464] Diameters with multiple entries such as "60 $\times$ 40 $\times$ 34" reflect that the body is not a perfect spheroid and that each of its dimensions have been measured well enough.

[465] Periods with negative values are retrograde.

[466] "?" refers to group assignments that are not considered sure yet.

[467] "Amalthea - definition of Amalthea in English from the Oxford dictionary" https://www.oxforddictionaries.com/definition/english/amalthea. OxfordDictionaries.com. Retrieved 20 January 2016.

[468] "Europa - definition of Europa in English from the Oxford dictionary" https://www.oxforddictionaries.com/definition/english/europa. OxfordDictionaries.com. Retrieved 20 January 2016.

[469] "Ganymede - definition of Ganymede in English from the Oxford dictionary" https://www.oxforddictionaries.com/definition/english/ganymede. OxfordDictionaries.com. Retrieved 20 January 2016.

[470] "Ganymede" https://www.merriam-webster.com/dictionary/Ganymede. *Merriam-Webster Dictionary*.

[471] File:Pioneer-10 jupiter moons.jpg

[472] Juno Approach Movie of Jupiter and the Galilean Moons https://www.missionjuno.swri.edu/media-gallery/jupiter-orbit-insertion?show=fig_577b4aae48b4964f5a8cd178&m=577b4aae48b4964f5a8cd178, NASA, July 2016

[473] http://home.dtm.ciw.edu/users/sheppard/satellites/jupsatdata.html

[474] http://home.dtm.ciw.edu/users/sheppard/satellites/

[475] https://web.archive.org/web/20110823204424/http://orinetz.com/planet/tourprog/jupitermoons.html

[476] http://alienworlds.southwales.ac.uk/jovianMoons.html

[477] http://solarsystem.nasa.gov/planets/jupiter/moons

[478] http://solarsystem.nasa.gov

[479] http://www.sfgate.com/news/article/43-more-moons-orbiting-Jupiter-2648039.php

[480] http://www.psrd.hawaii.edu/Archive/Archive-Jupiter.html

[481] http://www.psrd.hawaii.edu/index.html

[482] http://www.orbitsimulator.com/gravity/articles/joviansystem.html

[483] http://www.astronomycast.com/2007/10/episode-57-jupiters-moons/

[484] Drake, Stillman (1978). Galileo At Work. Chicago: University of Chicago Press.

[485]

[486] Zezong, Xi, "The Discovery of Jupiter's Satellite Made by Gan De 2000 years Before Galileo", *Chinese Physics* 2 (3) (1982): 664–67.

[487] *Cosimo* is the Italian form of the Greek name *Cosmas* itself deriving from *cosmos* (whence the neuter plural adjective *cosmica*). *Sidera* is the plural form of the Latin noun *sidus* "star, constellation".

[488] Howse, Derek. *Greenwich Time and the Discovery of the Longitude*. Oxford: Oxford University Press, 1980, 12.

[489] Computed using the IAU-MPC Satellites Ephemeris Service https://archive.is/20110520031937/http://www.minorplanetcenter.org/iau/NatSats/NaturalSatellites.html μ value

[490] Moons of Jupiter https://www.nasa.gov/sites/default/files/files/Moons_of_Jupiter_Lithograph.pdf *NASA*

[491] Computed from IAG Travaux 2001 http://www.hnsky.org/iau-iag.htm .

[492] Schenk, P. M.; Chapman, C. R.; Zahnle, K.; Moore, J. M.; *Chapter 18: Ages and Interiors: the Cratering Record of the Galilean Satellites*, in *Jupiter: The Planet, Satellites and Magnetosphere*, Cambridge University Press, 2004

[493] Jupiter is about 750 times brighter than Ganymede and about 2000 times brighter than Callisto. Ganymede: (5th root of 100)^(4.4 Ganymede APmag – (–2.8 Jup APmag)) = 758 Callisto: (5th root of 100)^(5.5 Callisto APmag – (–2.8 Jup APmag)) = 2089

[494] Jupiter near perihelion 2010-Sep-19: 656.7 (Callisto angular separation arcsec) – 24.9 (jup angular radius arcsec) = 631 arcsec = 10 arcmin

[495] http://strangepaths.com/observation-of-jupiter-moons-march-1613/2007/04/22/en/

[496] http://www.skyandtelescope.com/observing/objects/javascript/3307071.html

[497] "Lunar marriage may have given Jupiter a ring" https://www.newscientist.com/article/mg20527523.400-lunar-marriage-may-have-given-jupiter-a-ring.html, New Scientist, March 20, 2010, p. 16.

[498] The normal optical depth is the ratio between the total cross section of the ring's particles to the square area of the ring.

[499] The forward-scattered light is the light scattered at a small angle relative to solar light.

[500] The back-scattered light is the light scattered at an angle close to 180° relative to solar light.

[501] ^ It should be compared with approximately 1700 km² total cross section of Metis and Adrastea.

[502] Lorentz resonance is a resonance between particle's orbital motion and rotation of planetary magnetosphere, when the ratio of their periods is a rational number.

[503] The thickness of the gossamer rings is defined here as the distance between peaks of brightness at their top and bottom edges.

[504] http://nssdc.gsfc.nasa.gov/planetary/factsheet/jupringfact.html

[505] http://solarsystem.nasa.gov/planets/profile.cfm?Object=Jupiter&Display=Rings

[506] http://solarsystem.nasa.gov

[507] https://web.archive.org/web/20060206090452/http://spaceprojects.arc.nasa.gov/Space_Projects/pioneer/PNhome.html

[508] http://voyager.jpl.nasa.gov

[509] http://www2.jpl.nasa.gov/galileo

[510] https://web.archive.org/web/20070426162116/http://saturn.jpl.nasa.gov/home/index.cfm

[511] http://pluto.jhuapl.edu

[512] http://pds-rings.seti.org/jupiter/

[513] http://planetarynames.wr.usgs.gov/Page/Rings#jupiter

[514] http://planetarynames.wr.usgs.gov

Article Sources and Contributors

The sources listed for each article provide more detailed licensing information including the copyright status, the copyright owner, and the license conditions.

Jupiter *Source:* https://en.wikipedia.org/w/index.php?oldid=855736055 *License:* Creative Commons Attribution-Share Alike 3.0 *Contributors:* *thing goes, A2soup, Adpete, Aeonx, Agmartin, Aledownload, Alpinu, Amccann421, Anand2202, AnimeJanai, AnotherPoint, Arado, Aria1561, BD2412, BatteryIncluded, Bd7941a, Begoon, Bender235, Bgwhite, Bigdaddybrabantio, Brandmeister, Brraflor, Cavrdg, Ceasars Salad, Checkingfax, ChiZeroOne, Chricho, CircleAdrian, ClueBot NG, Cobblet, Crcwiki, CubeSat4U, CuriousMind01, DN-boards1, DSmurf, David spector, DavideVeloria88, Dawnseeker2000, Devinbeth, Devinjackson199, Dewritech, Dhtwiki, Dilidor, DocWatson42, Dondervogel 2, DooFi, Double sharp, Doublesuited, DrKay, Dratman, Drbogdan, EP111, Edulovers, EeeveeeFrost, Elinor Rajka, Enelson, Eric Kvaalen, EuroAgurbash, FlightTime, Frmorrison, G0mx, Gcanyon, Gorthian, Gouzmalix, Graeme Bartlett, Hadron137, HalloweenNight, Headbomb, Hillbillyholiday, Huntster, Huritisho, Iggy the Swan, Isambard Kingdom, JDAWiseman, Jarble, Jennica, Jidanni, Jmrowland, Jni, Joefromrandb, Jon Kolbert, JorisEnter, JorisvS, Joss24, Julien1978, Karn, Kazvorpal, Kheider, Knochen, Krj373, Kwamikagami, Lambiam, Largoplazo, Lithopsian, LlywelynII, Lockesdonkey, Lylahearts, MFH, Maczkopeti, Magioladitis, Marconi sparks, MarcusGR, MarioProtIV, Mark the train, MartinZ, Materialscientist, Melikamp, Mikhail Ryazanov, Mitchumch, Mount2010, Mr Stephen, NPalgan2, Nagualdesign, Nardog, Narky Blert, Nebulousness, Nihiltres, Omnipaedista, Onel5969, Orion415, Paintspot, Pasachoff, PcPrincipal, Pdebee, PhilipTerryGraham, Phyzome, Piperh, PlanetUser, Plastikspork, Randy Kryn, Red Director, Remotelysensed, Rfassbind, RileyBugz, Rjwilmsi, Rod57, Rothorpe, Sachinvenga, SandyGeorgia, Saros136, Saurusaurus, ScientistMr, Scottperry, SenseiAC, Sholokhov, Sizeofint, Spartan - 117, StewartIM, Summerdrought, TAnthony, Tamfang, Tarunuee, Tglaisyer, The Transhumanist, TheWhistleGag, ThiefOfBagdad, Thingg, Tillman, Tom.Reding, Tony1, U-95, User-duck, Vgy7ujm, Vmavanti, Voello, W like wiki, WereSpielChequers, White whirlwind, WolfmanSF, Xactnorge, Xover, YuRi YuZi, Zedshort, Zekelayla, Zenedits, Ρουθραμιώτης, גולן של אמא 1

Jupiter mass *Source:* https://en.wikipedia.org/w/index.php?oldid=844647885 *License:* Creative Commons Attribution-Share Alike 3.0 *Contributors:* -Paul-, Anders Sandberg, Anonymous Dissident, Beepy, Bender235, Bert depoorter, Bgwhite, BotKung, Chaos syndrome, Citation bot 1, ClueBot NG, Colin.campbell.27, Dcirovic, Drbogdan, Esmera en, Excirial, Gap9551, Giraffedata, Gyrobo, Hadron137, HarryAlffa, Hdjensofjfnen, Hibernian, Interstellar Man, Jkl, Jolielegal, JorisvS, Kheider, Largoplazo, Magioladitis, Materialscientist, Mike s, Mimihitam, Paine Ellsworth, Physchim62, PlanetStar, RJHall, Remember, Rfassbind, Rjwilmsi, Roentgenium111, Ruslik0, Sardanaphalus, Serendipodous, Skittleys, SkyFlubbler, Smeagol 17, Sonicology, Spacepotato, TAKASUGI Shinji, TJRC, Tom.Reding, Tomchiukc, Wiac, Zginder, عباد مجاهد ديرانية, 28 anonymous edits 33

Atmosphere of Jupiter *Source:* https://en.wikipedia.org/w/index.php?oldid=852545541 *License:* Creative Commons Attribution-Share Alike 3.0 *Contributors:* 4.0gpacollegegrad, 4pq1injbok, A. Parrot, Admdata13, Alex Nico, Alexander Davronov, Ariadacapo, BD2412, Bender235, Bill.D Nguyen, Bronze2018, CharlesC, Chris goulet, ClueBot NG, Cnwilliams, Corymchapman, DJFission, Danim, Diannaa, Drbogdan, Ed Poor, Edward, El C, Elk Salmon, Faizan, Finnusertop, Frosty, Gadget850, Gilliam, Glane23, Gob Lofa, Grinder34, Harish R Manikandan, Headbomb, Hellbus, Heron, Hitro-Milanese, Huntster, I dream of horses, Iggymwangi, Iksnyrk, Int21h, Jessegalebaker, Jim.henderson, Jmencisom, John, JorisvS, Jupiter-USA, K9re11, KapteynCook, KarenAplin, Katieh5584, Kazvorpal, Kheider, KingMunch, Kitch, Klilidiplomus, Labtek00, Lagrange613, Libcard, Lieninger, Magioladitis, Maniesansdelire, Materialscientist, Mathonius, Mattster3517, Max.kit, Medeis, Mhh1000, Moe Epsilon, Mogism, Mr Stephen, Mz7, Nasnema, Newone, Newyork1501, Nimesh Mistry, Ninney, Nyttend, Originalwana, PINEKEV10, Peace Makes Plenty, Philip Trueman, PhilipTerryGraham, PlanetStar, PlanetUser, Quebec99, RA0808, Racerx11, Rain drop 45, RandomLittleHelper, RhinoMind, Rjwilmsi, RobertG, RockMagnetist (DCO visiting scholar), Rursus, Ruslik0, Sandvich18, ScottSteiner, Serendipodous, Sidelight12, Skarebo, Slazenger, Solarra, Spicemix, Spinningspark, Steel1943, Stephenb, Szczureq, Tbhotch, Tdadamemd, Tentinator, The High Fin Sperm Whale, TheAustinMan, Thirdright, Tmangray, Toddst1, Tom.Reding, Trappist the monk, Tycho Magnetic Anomaly-1, Victor falk, Vsmith, Wavelength, Wiki-Ilya, William2001, Wingedsubmariner, WolfmanSF, Zojj, Zyksnowy, 179 anonymous edits 37

Magnetosphere of Jupiter *Source:* https://en.wikipedia.org/w/index.php?oldid=855683608 *License:* Creative Commons Attribution-Share Alike 3.0 *Contributors:* -Strogoff-, Aarghdvaark, Adrian J. Hunter, Aldebaran66, Alex Cohn, Anthonyhcole, Ariel A Delgado, Art LaPella, Ashorocetus, BatteryIncluded, Bender235, Borrow-188, CardinalDan, Chris the speller, Citation bot 1, Clinamental, ClueBot NG, CommonsDelinker, Courcelles, Cryptic C62, Dabomb87, Dcirovic, Deneb in Cygnus, DocWatson42, Dodshe, Doradus, Double sharp, DrKay, Drbogdan, Earthandmoon, East718, Eluchil404, Eyreland, Fcrary, Fotaun, Fountains of Bryn Mawr, Gadget850, Gene Nygaard, GorillaWarfare, GrahamHardy, GreekAlexander, HJ Mitchell, HamburgerRadio, Haruth, Headbomb, Huntster, Jairodz, JamesBWatson, Jappalang, John of Reading, Jonny Hansson, JorisvS, Jovianeye, Julesd, Jupiter-4, Kethrus, Kuralyov, Lambiam, Lancevortex, LeighAlexH, Mark Arsten, Materialscientist, Metilisopropilisergamida, MfortyoneA, Mgiganteus1, Mhshaik, Mike Peel, Mikeo, Mild Bill Hiccup, Mnmngb, Monolithica, Multixfer, Myasuda, Nergaal, Nihiltres, Nono64, Ntsimp, Ohms law, Otr500, Pauli133, Piledhigheranddeeper, Plant's Strider, Quibik, Quisquillian, R'n'B, RJHall, RexxS, Rich Farmbrough, Richfj, Rjwilmsi, Rock4arolla, Ruslik0, SandyGeorgia, Sardur, Serendipodous, Silverblaze4575, Spiral5800, Spotty11222, Steel1943, Suffusion of Yellow, Telekenesis, The profit mohammed, ThiefOfBagdad, Thumperward, Timbliboo, TimothyRias, Titoxd, Tom.Reding, Trevayne08, Tycho Magnetic Anomaly-1, Wavelength, Welsh, Whoop whoop pull up, Wikiain, Wjejskenewr, WolfmanSF, Xeeron, 52 anonymous edits 75

Exploration of Jupiter *Source:* https://en.wikipedia.org/w/index.php?oldid=856260690 *License:* Creative Commons Attribution-Share Alike 3.0 *Contributors:* A Karley, AManWithNoPlan, Aldebaran66, AlexTingle, Alfie Gandon, Alpha3031, AntHerder, Anythingcouldhappen, Apollo The Logician, Arjayay, Article editor, Auric, Barsha kanungo, BatteryIncluded, Bellerophon, Bender235, Bgwhite, Blaisorblade, CLCStudent, Charvest, Citation bot 1, Citizen Canine, ClueBot NG, Cmglee, Cowik, Cuddy Wifter, Czolgolz, DPBT1, Danrob0910, DataWraith, Dawnseeker2000, Dcirovic, Drbogdan, Elizabeth Linden Rahway, Emilsson50, Eric-Wester, Eyreland, Fartherred, Flat Out, Fotaun, Fraggle81, Gdiazc, Geometry guy, Gilliam, Gob Lofa, Gustavobaldo, GünniX, Hans Dunkelberg, Headbomb, Hmains, Hmainsbot1, Huritisho, I7s, ILikeTurtles21, Ikokki, Ilmari Karonen, Inhighspeed, Iridescent, Ixfd64, J 1982, JFG, Jacobmayle, Jazzyjaffa, Jim1138, Jimknut, John, Jolielegal, JorisvS, Kees08, Kheider, Kitch, Knochen, Koavf, Ling.Nut, Little Mountain 5, MBK004, MKUV, Mabandalone, Marasama, Mark Foskey, Marshasaleins, Materialscientist, Matsaball, Mild Bill Hiccup, Mx. Granger, Nergaal, OcarinaOfTime, Phoenix7777, Pladask, PlanetUser, Plant's Strider, Quebec99, Qwidjib0, Randy Kryn, Reywas92, RhinoMind, Rich Farmbrough, Ricklaman, Rjwilmsi, Rreagan007, Ruslik0, Serendipodous, Serols, Sethton, Snjón, Stone, Stumpfian, Surajt88, Svensvensonson, Tabletop, Telekenesis, Tetra quark, Thor Dockweiler, Titodutta, Tktktk, Tom.Reding, V1adis1av, Varkman, Whoop whoop pull up, ZackTheJack, 95 anonymous edits 103

Moons of Jupiter *Source:* https://en.wikipedia.org/w/index.php?oldid=856109663 *License:* Creative Commons Attribution-Share Alike 3.0 *Contributors:* Abductive, Ahonc, Andrewjohnbayles, Anomalocaris, Archon 2488, ArnoldReinhold, Astronautics~enwiki, Bedzior, Bowling is life, C.Fred, CLCStudent, CamHat000099, Carmenbonita, ClueBot NG, Codwiki, Coffeeandcrumbs, DVdm, Dag330, David.moreno72, Dcirovic, Dhtwiki, Dilidor, Dinnypaul, DinoSlider, Double sharp, Dreigorich, Ehrenkater, Elk Salmon, Emori89, Exoplanetaryscience, Gap9551, Geoffrey.landis, Gustavo1100, Hamiltondaniel, Headbomb, Hl, HolyT, Iggy the Swan, Ivancurtisivancurtis, J 1982, J. 'mach' wust, JDAWiseman, JEH, Jim1138, JorisvS, Juice987, KH-1, KNHaw, Kwamikagami, Laoris, Leafmealone1337, Lucca360mlg, LukeSurl, Maczkopeti, Malihahasan2, Materialscientist, Melonkelon, Michael hefler, Milochicken, Mong2266, Nagualdesign, Nardog, Nergaal, Ninney, Northamerica1000, Orion415, Oshwah, PackMecEng, PhilipTerryGraham, PlanetUser, Renerpho, Richard3120, Rjwilmsi, Roentgenium111, Roger de Lauria, Ruslik0, Rwv37, Seattlebread, SemiHypercube, Serendipodous, Serols, Shellwood, Shona0308, Soerfm, Sonicwave32, Steel1943, StewartIM, Stikkyy, SusanLesch, Szabi, Tglaisyer, The NMI User, TheWikipediaPerson, Tianl3, Tommy-Boy, Tonyhallmailbox, TwoTwoHello, Ugog Nizdast, Vanamonde93, Vimalkalyan, Waggers, Whoisjohngalt, Yoshi24517, Zaphody3k, Zedshort, Zyxok, İnternion, גולן של אמא, 140 anonymous edits 119

Galilean moons *Source:* https://en.wikipedia.org/w/index.php?oldid=852882485 *License:* Creative Commons Attribution-Share Alike 3.0 *Contributors:* Al Legorhythm, Alex102322, AlexanderHaas, Awilley, Bautsch, Bender235, Bongwarrior, BoxofPresents, CAPTAIN RAJU, CCthepie, Cannolis, Chiswick Chap, Chris Capoccia, Chris the speller, Citation bot 1, Classicwiki, ClueBot NG, Cmglee, Colby890, Colonies Chris, Dawn Bard, Dawnseeker2000, Dcirovic, Djaquishy, Doublesuited, Drbogdan, El C, Emily Verschoor, Epbr123, EvenGreenerFish, Exoplanetaryscience, Fjeinca, Flyer22 Reborn, Fornever700, Fotaun, Gob Lofa, Graham87, Guy vandegrift, Hans Dunkelberg, Haselbrunner278, Headbomb, Howicus, Huritisho, I am One of Many, Ipwnu4ev1, IronGargoyle, J 1982, JasonAQuest, Jcpag2012, Jim1138, Jmencisom, Jncraton, John, John of Reading, JorisvS, Jsolinsky, KNHaw, Kheider, Kjhrbkwbfv, Kwamikagami, KylieTastic, Llewes, Lord Roem, Lugia2453, MKUV, Macaddct1984, Magioladitis, Mccapra, Mejor Los Indios, Melonkelon, Messier8, Michael hefler, Mike Peel, Misty MH, MrBark, N0n3up, Natuur12, Neo., Ninney, Nwbeeson, Nyttend, Oberlinjoe, Onel5969, PJtP, Petter Per, Pikamander2, Pinethicket, Plant's Strider, Poltair, Polyamorph, Quebec99, Randy Kryn, Rhododendrites, Rjwilmsi, Rowellcf, Ruslik0, SQGibbon, Sam Sailor, Serendipodous, Shellwood, SheriffIsInTown, Simplexity22, Sjö, Smalljim, Soerfm, Solomonfromfinland, Sonicyouth86, Steve03Mills, StewartIM, Stvltvs, TAnthony, Teammm, TeigeRyan, Tetra quark, Tewapack, Tglaisyer, Thanatos666, TheOtherUnknown, Thewinrat, Thony C., Tianl3, Tom.Reding, Tryphiodorus, Tutelary, Twinsday, Tyler97657, Vsmith, WadeSimMiser, Wayne Slam, Wiae, WolfmanSF, Wtmitchell, Yamaguchi先生, Yunshui, ZelyaV, Zupotachyon, Δουλοιεσμεν, 159 anonymous edits 141

Rings of Jupiter *Source:* https://en.wikipedia.org/w/index.php?oldid=856092671 *License:* Creative Commons Attribution-Share Alike 3.0 *Contributors:* Acetotyce, Ahmedkandiel, Anaerobe2738, Andplus, Art LaPella, Arx Fortis, Asenine, Attilios, BOTarate, Bender235, Bgpaulus, BlueMoonlet, Bri, CAPTAIN RAJU, CWii, Cacophony, Catgut, Chaser, Chesnok, Citation bot 1, ClueBot NG, ComicsAreJustAllRight, CommonsDelinker, Cshell13, Dbo789, Dedalus, Deviator13, Double sharp, Dougofborg, Dreamafter, Earthandmoon, Elberth 00001939, Ember of Light, Episcophagus, Exoplanetaryscience, Fish and karate, Fotaun, Fraggle81, Frosted14, Gap9551, Gene Nygaard, Giraffedata, GrahamHardy, GregorB, Guardianrule, Gutzky, Headbomb, Hellbus, Henrykus, Huritisho, Iomodian, ItsZippy, Ixfd64, J 1982, J.delanoy, JForget, JaGa, Jerome Kelly, Jnestorius, John of Reading, Joijfeoi, JorisvS, K6ka, Kalisitahi, Katydidit, Keta, Kethrus, Kimchi.sg, Knvs99, Koavf, Kozuch, Kwamikagami, Ladyofwisdom, Lanthanum-138, Lightmouse, Lugia2453,

Image Sources, Licenses and Contributors

The sources listed for each image provide more detailed licensing information including the copyright status, the copyright owner, and the license conditions.

License

Index

Extraterrestrial oceans, 127

Ferdinando II de Medici, Grand Duke of Tuscany, 145
Field aligned current, 86
Field strength, 75
Fieseler, 181
File:Jupiter from Voyager 1 PIA02855 max quality.ogv, 11
File:Pioneer-10 jupiter moons.jpg, 184
Filipino people, 58
Fireball (meteor), 30
Five elements (Chinese philosophy), 17
Flattening, 2
Flecther2010, 177
Fluid mechanics, 51
Flux, 166, 169
Flux tube, 83
Folk etymology, 19
Foot per second squared, 2
Foreshortening, 45
Formation and evolution of the Solar System, 42
France, 147
Frank Drake, 21
Friction, 26
Full moon, 78, 158
Fu star, 17

Galilean moon, 19, 23, 108, 130, 131
Galilean moons, 4, 16, 26, 27, 77, 105, 112, 119, 120, 122, 123, 126, 129, 139, **141**
Galilean satellites, 89
Galileo Galilei, 4, 19, 27, 119, 122, 130, 141, 143
Galileo orbiter, 67
Galileo Probe, 23, 39, 41, 44, 50, 63, 64
Galileo spacecraft, 23, 83, 104, 113, 139
Galileo (spacecraft), 4, 41, 62, 91, 95, 103, 112, 114, 161, 165, 169, 171, 172, 174
Gan De, 19, 122, 144
Ganymede (moon), 4, 26, 87, 91, 107, 112, 114, 115, 117, 119, 122, 124–126, 130, 141, 142, 144, 146, 148
Ganymede (mythology), 152
Gas constant, 67
Gas giant, 3, 41
Gas torus, 76
Gauss (unit), 13, 75, 77, 92
Gemini Observatory, 60
Geocentric, 19
Geocentric model, 141, 143
Geocentric orbit, 105
Geographical pole, 2
Geographic pole, 15
Geometric albedo, 3

Geosynchronous Satellite Launch Vehicle Mark III, 105
Germane, 42
Germanic paganism, 18
G-force, 2, 113
Gian Domenico Cassini, 54
Giant planet, 3, 12
Gigameter, 128
Gigametre, 126
Giovanni Alfonso Borelli, 20
Giovanni Battista Hodierna, 145
Giovanni Battista Riccioli, 54
Giovanni Cassini, 65
Giovanni Domenico Cassini, 20, 30, 147
GJ 229B, 36
Gliese 229B, 33
Go, 177
Goddard Institute for Space Studies, 110
Goddard Space Flight Center, 60
God (male deity), 17
Grammatical gender, 184
Grand tack hypothesis, 4
Graney, 176, 177
Gravitational constant, 35
Gravitational slingshot, 4, 22, 105
Gravitational well, 116
Gravity, 26, 28
Gravity field, 114
Gravity wave, 40
Gravity well, 29
Great Dark Spot, 57
Great Red Spot, 4, 11, 12, 16, 37, 65, 108, 110, 153, 175
Greek language, 17
Greek mythology, 17, 119, 152
Guernsey, 184
Guido Bonatti, 17
Guru, 17
Gyroscope, 95, 106

Hadley cell, 44
Hammel, 177
Harpalyke (moon), 134
HD 209458 b, 5
Heat flux, 51
Heat of vaporization, 40
Hectometer, 88
Hegemone (moon), 137
Heimpel2005, 176
Helike (moon), 133
Heliocentrism, 20
Heliosphere, 76
Helium, 3, 37, 41, 116
Helium-3, 41, 116
Helium-4, 41
Hera, 149